221
Advances in Polymer Science

Advances in Polymer Science

Recently Published and Forthcoming Volumes

Advanced Computer Simulation Approaches for Soft Matter Sciences III

Volume Editors: Christian Holm · Kurt Kremer

With contributions by

P.G. Bolhuis · C. Dellago · B. Dünweg · G. Gompper
T. Ihle · D.M. Kroll · A.J.C. Ladd · R.G. Winkler

 Springer

The series *Advances in Polymer Science* presents critical reviews of the present and future trends in polymer and biopolymer science including chemistry, physical chemistry, physics and material science. It is adressed to all scientists at universities and in industry who wish to keep abreast of advances in the topics covered.

As a rule, contributions are specially commissioned. The editors and publishers will, however, always be pleased to receive suggestions and supplementary information. Papers are accepted for *Advances in Polymer Science* in English.

In references *Advances in Polymer Science* is abbreviated *Adv Polym Sci* and is cited as a journal.

Springer WWW home page: springer.com
Visit the APS content at springerlink.com

ISBN: 978-3-540-87705-9   e-ISBN: 978-3-540-87706-6
DOI: 10.1007/978-3-540-87706-6

Advances in Polymer Science ISSN 0065-3195

Library of Congress Control Number: 2008941064

Cover design: WMXDesign GmbH, Heidelberg

Typesetting: SPi

Printed on acid-free paper

springer.com

Advances in Polymer Science
Also Available Electronically

For all customers who have a standing order to Advances in Polymer Science, we offer the electronic version via SpringerLink free of charge. Please contact your librarian who can receive a password or free access to the full articles by registering at:

springerlink.com

If you do not have a subscription, you can still view the tables of contents of the volumes and the abstract of each article by going to the SpringerLink Home page, clicking on "Browse by Online Libraries", then "Chemical Sciences", and finally choose Advances in Polymer Science.

You will find information about the

– Editorial Board
– Aims and Scope
– Instructions for Authors
– Sample Contribution

at springer.com using the search function.

Color figures are published in full color within the electronic version on SpringerLink.

Preface

"Soft matter" is nowadays used to describe an increasingly important class of materials that encompasses polymers, liquid crystals, molecular assemblies building hierarchical structures, organic-inorganic hybrids, and the whole area of colloidal science. Common to all is that fluctuations, and thus the thermal energy $k_B T$ and entropy, play an important role. "Soft" then means that these materials are in a state of matter that is neither a simple liquid nor a hard solid of the type studied in hard condensed matter, hence sometimes many types of soft matter are also named "complex fluids."

Soft matter, either of synthetic or biological origin, has been a subject of physical and chemical research since the early finding of Staudinger that long chain molecules exist. From then on, synthetic chemistry as well as physical characterization underwent an enormous development. One of the outcomes is the abundant presence of polymeric materials in our everyday life. Nowadays, methods developed for synthetic polymers are being more and more applied to biological soft matter. The link between modern biophysics and soft matter physics is quite close in many respects. This also means that the focus of research has moved from simple homopolymers to more complex structures, such as branched objects, heteropolymers (random copolymers, proteins), polyelectrolytes, amphiphiles and so on. While basic questions concerning morphology, dynamics, and rheology are still a matter of intense research, additional, more advanced topics are also being tackled, for example the link between structure and function or non-equilibrium aspects.

For many years there have been attempts to understand these systems thoroughly using theoretical concepts. Beginning with the early work of Flory, simplified models were studied, which were able to explain certain generic/universal aspects but failed to provide a solid theoretical basis for this universal behavior. It was then up to the seminal works of de Gennes and Edwards to provide a link between the statistical mechanics of phase transitions (critical phenomena) and polymer chain conformations. This link to the modern concepts of theoretical physics provided huge momentum for the field, which shaped many theoretical schools and formed the basis for modern soft matter physics. Despite all these developments, soft matter theory is still an active and growing research field. Due to the high degree of complexity of the problems it is not surprising that analytical theory can only treat highly idealized and simplified models. Consequently, with the availability of computers, problems in polymer science were among the first to be tackled by simulations. Even now, the problem of an isolated self-avoiding walk cannot be solved exactly in three dimensions. As early as 1954, Hammersley and Morton, and Rosenbluth and Rosenbluth tried to overcome the related attrition problem in growing self-avoiding walks by introducing "inversely restricted sampling." In addition, basic multichain features (such as the noncrossability of chains) are hard to deal with analytically and can only be included properly by a simulation approach. Thus, with the rising availability of computing power, simulation methods began to play an increasingly important role in soft matter research. Computing power is, however, only one aspect. Even more important has been the development of advanced numerical methods and highly optimized programs. Very different areas, ranging from quantum chemistry studying molecules on the sub-Ångstrøm level all the way to macroscopic fluid dynamics, have to come together and offer a unique set of research opportunities. Over the years, the role of computer simulations has gone beyond the traditional aspect of checking approximative solutions of analytical models and bridging the gap between experiments and theory. They are now an independent, in some cases even predictive, tool in materials research, for example for complex molecular assemblies or specific rheological problems.

It is the purpose of this small series of volumes in *Advances in Polymer Sciences* to provide an overview of the latest developments in the field. For this, internationally renowned experts review recent work in the general area of soft matter simulations. The third volume contains three contributions. The first two chapters review several coarse-grained methods to include the effects of hydrodynamics in mesoscopic particle simulations that use an implicit solvent, whereas the last chapter deals with advanced sampling methods to study rare events.

The first two contributions deal with methods or systems where hydrodynamic interactions play a dominant role. Studying coarse-grained mesoscopic systems, hydrodynamic interactions are unimportant for static properties in equilibrium. However, the inclusion of hydrodynamic effects becomes indispensable for all problems of dynamics of solutions in bulk or under confinement, especially when it comes to flow-induced structure formation. This would automatically be achieved by a standard molecular dynamics simulation, which takes full account of the solvent molecules. This, however, is only feasible in some very exceptional cases, even for the upcoming computer generation, and is still applicable to only very small systems. Because of that, solvent-free methods play a very important role and have been improved significantly over the last few years. In the first contribution G. Gompper, T. Ihle, D.M. Kroll, and R.G. Winkler focus on an algorithm that was initially proposed by Malevanets and Kapral in 1999, and is now called multiparticle collision dynamics (MPC) or stochastic rotation dynamics (SRD). The method consists of alternating streaming and collision steps in an ensemble of point particles that locally conserve mass, momentum, and energy. The second contribution by B. Dünweg and A.J.C. Ladd reviews in depth the standard D3Q19 lattice-Boltzmann model and extensions thereof. Here the Boltzmann equation is solved on a grid, where the fluid velocities are stored, employing mass and momentum conservation. The authors discuss in depth the "fluctuating" lattice-Boltzmann algorithm, followed by a detailed discussion of complementary methods for the coupling of solvent and solute. Both presented methods consistently couple full hydrodynamic interactions and thermal fluctuations and, since they deal with complementary methods, give an excellent comprehensive overview over the field. Both contributions also conclude with examples in which the methods are applied to soft matter systems such as colloidal suspensions and polymer solutions.

In the third contribution, C. Dellago and P. Bolhuis review several recently developed methods for studying rare-event transitions, which are important in understanding molecular processes such as nucleation events, chemical reactions transport phenomena in liquids and solids, or slow processes such as protein folding. Such transition events are rare because the stable basins are separated from each other by high free-energy barriers of either potential energy, entropic, or combined origin. Several methods have been proposed to speed up the sampling of these transitions, like metadynamics, the finite temperature string method, forward flux sampling, and others. The authors cover in depth the transition path sampling methodology to which they have both added important contributions.

We are confident that this collection of reviews will be a very useful guide to interested scientists and advanced students, and it also provides detailed background information for experienced researchers in the field.

Mainz, Autumn 2008   *C. Holm, K. Kremer*

Contents

Contents of

Adv Polym Sci 221: 1–87
DOI:10.1007/12_2008_5
© Springer-Verlag Berlin Heidelberg 2008

Multi-Particle Collision Dynamics: A Particle-Based Mesoscale Simulation Approach to the Hydrodynamics of Complex Fluids

G. Gompper, T. Ihle, D.M. Kroll, and R.G. Winkler

Abstract In this review, we describe and analyze a mesoscale simulation method for fluid flow, which was introduced by Malevanets and Kapral in 1999, and is now called multi-particle collision dynamics (MPC) or stochastic rotation dynamics (SRD). The method consists of alternating streaming and collision steps in an ensemble of point particles. The multi-particle collisions are performed by grouping particles in collision cells, and mass, momentum, and energy are locally conserved. This simulation technique captures both full hydrodynamic interactions and thermal fluctuations. The first part of the review begins with a description of several widely used MPC algorithms and then discusses important features of the original SRD algorithm and frequently used variations. Two complementary approaches for deriving the hydrodynamic equations and evaluating the transport coefficients are reviewed. It is then shown how MPC algorithms can be generalized to model non-ideal fluids, and binary mixtures with a consolute point. The importance of angular-momentum conservation for systems like phase-separated liquids with different viscosities is discussed. The second part of the review describes a number of recent applications of MPC algorithms to study colloid and polymer dynamics, the behavior of vesicles and cells in hydrodynamic flows, and the dynamics of viscoelastic fluids.

Keywords Binary fluid mixtures, Colloids, Complex fluids, Hydrodynamics, Mesoscale simulation techniques, Microemulsions, Polymers, Red blood cells, Vesicles, Viscoelastic fluids

G. Gompper (✉) and R.G. Winkler
Theoretical Soft Matter and Biophysics, Institut für Festkörperforschung, Forschungszentrum Jülich, 52425 Jülich, Germany
e-mail: g.gompper@fz-juelich.de; r.winkler@fz-juelich.de

T. Ihle and D.M. Kroll
Department of Physics, North Dakota State University, Fargo, ND 58108-6050, USA
e-mail: thomas.ihle@ndsu.edu; daniel.kroll@ndsu.edu

Contents

1 Introduction

"Soft Matter" is a relatively new field of research that encompasses traditional complex fluids such as amphiphilic mixtures, colloidal suspensions, and polymer solutions, as well as a wide range of phenomena including chemically reactive flows (combustion), the fluid dynamics of self-propelled objects, and the visco-elastic behavior of networks in cells. One characteristic feature of all these systems is that phenomena of interest typically occur on mesoscopic length-scales – ranging from nano- to micrometers – and at energy scales comparable to the thermal energy $k_B T$.

Because of the complexity of these systems, simulations have played a particularly important role in soft matter research. These systems are challenging for conventional simulation techniques due to the presence of disparate time, length, and energy scales. Biological systems present additional challenges because they are often far from equilibrium and are driven by strong spatially and temporally varying forces. The modeling of these systems often requires the use of "coarse-grained" or mesoscopic approaches that mimic the behavior of atomistic systems on the length scales of interest. The goal is to incorporate the essential features of the microscopic physics in models which are computationally efficient and are easily implemented in complex geometries and on parallel computers, and can be used to predict emergent properties, test physical theories, and provide feedback for the design and analysis of experiments and industrial applications.

In many situations, a simple continuum description based on the Navier–Stokes equation is not sufficient, since molecular-level details – including thermal fluctuations – play a central role in determining the dynamic behavior. A key issue is to resolve the interplay between thermal fluctuations, hydrodynamic interactions, and spatio-temporally varying forces. One well-known example of such systems are microemulsions – a dynamic bicontinuous network of intertwined mesoscopic patches of oil and water – where thermal fluctuations play a central role in *creating* this phase. Other examples include flexible polymers in solution, where the coil state and stretching elasticity are due to the large configurational entropy. On the other hand, atomistic molecular dynamics simulations retain too many microscopic degrees of freedom, consequently requiring very small time steps in order to resolve the high frequency modes. This makes it impossible to study long timescale behavior such as self-assembly and other mesoscale phenomena.

In order to overcome these difficulties, considerable effort has been devoted to the development of mesoscale simulation methods such as Dissipative Particle Dynamics [1–3], Lattice-Boltzmann [4–6], and Direct Simulation Monte Carlo [7–9]. The common approach of all these methods is to "average out" irrelevant microscopic details in order to achieve high computational efficiency while keeping the essential features of the microscopic physics on the length scales of interest. Applying these ideas to suspensions leads to a simplified, coarse-grained description of the solvent degrees of freedom, in which embedded macromolecules such as polymers are treated by conventional molecular dynamics simulations.

All these approaches are essentially alternative ways of solving the Navier–Stokes equation and its generalizations. This is because the hydrodynamic equations are expressions for the local conservation laws of mass, momentum, and energy, complemented by constitutive relations which reflect some aspects of the microscopic details. Frisch et al. [10] demonstrated that discrete algorithms can be constructed which recover the Navier–Stokes equation in the continuum limit as long as these conservation laws are obeyed and space is discretized in a sufficiently symmetric manner.

The first model of this type was a cellular automaton, called the Lattice-Gas-Automaton (LG). The algorithm consists of particles which jump between nodes of a regular lattice at discrete time intervals. Collisions occur when more than one particle jumps to the same node, and collision rules are chosen which impose mass and momentum conservation. The Lattice-Boltzmann method (LB) – which follows the evolution of the single-particle probability distribution at each node – was a natural generalization of this approach. LB solves the Boltzmann equation on a lattice with a small set of discrete velocities determined by the lattice structure. The price for obtaining this efficiency is numerical instability in certain parameter ranges. Furthermore, as originally formulated, LB did not contain any thermal fluctuations. It became clear only very recently (and only for simple liquids) how to restore fluctuations by introducing additional noise terms to the algorithm [11].

Except for conservation laws and symmetry requirements, there are relatively few constraints on the structure of mesoscale algorithms. However, the constitutive relations and the transport coefficients depend on the details of the algorithm, so that the temperature and density dependencies of the transport coefficients can be quite different from those of real gases or liquids. However, this is not a problem as long as the *functional form* of the resulting hydrodynamic equations is correct. The mapping to real systems is achieved by tuning the relevant characteristic numbers, such as the Reynolds and Peclet numbers [12, 13], to those of a given experiment. When it is not possible to match all characteristic numbers, one concentrates on those which are of order unity, since this indicates that there is a delicate balance between two effects which need to be reproduced by the simulation. On occasion, this can be difficult, since changing one internal parameter, such as the mean free path, usually affects all transport coefficients in different ways, and it may happen that a given mesoscale algorithm is not at all suited for a given application [14–17].

In this review we focus on the development and application of a particle-based mesoscopic simulation technique which was recently introduced by Malevanets and Kapral [18, 19]. The algorithm, which consists of discrete streaming and collision steps, shares many features with Bird's Direct Simulation Monte Carlo (DSMC) approach [7]. Collisions occur at fixed discrete time intervals, and although space is discretized into cells to define the multi-particle collision environment, both the spatial coordinates and the velocities of the particles are continuous variables. Because of this, the algorithm exhibits unconditional numerical stability and has an H-theorem [18,20]. In this review, we will use the name multi-particle collision dynamics (MPC) to refer to this class of algorithms. In the original and most widely used version of MPC, collisions consist of a stochastic rotation of the relative velocities of the particles in a collision cell. We will refer to this algorithm as stochastic rotation dynamics (SRD) in the following.

One important feature of MPC algorithms is that the dynamics is well-defined for an arbitrary time step, Δt. In contrast to methods such as molecular dynamics simulations (MD) or dissipative particle dynamics (DPD), which approximate the continuous-time dynamics of a system, the time step does not have to be small. MPC *defines* a discrete-time dynamics which has been shown to yield the correct long-time hydrodynamics; one consequence of the discrete dynamics is that the transport coefficients depend explicitly on Δt. In fact, this freedom can be used to tune the Schmidt number, Sc [15]; keeping all other parameters fixed, decreasing Δt leads to an increase in Sc. For small time steps, Sc is larger than unity (as in a dense fluid), while for large time steps, Sc is of order unity, as in a gas.

Because of its simplicity, SRD can be considered an "Ising model" for hydrodynamics, since it is Galilean invariant (when a random grid shift of the collision cells is performed before each collision step [21]) and incorporates all the essential dynamical properties in an algorithm which is remarkably easy to analyze. In addition to the conservation of momentum and mass, SRD also locally conserves energy, which enables simulations in the microcanonical ensemble. It also fully incorporates both thermal fluctuations and hydrodynamic interactions. Other more established methods, such as Brownian Dynamics (BD) can also be augmented to include hydrodynamic interactions. However, the additional computational costs are often prohibitive [22, 23]. In addition, hydrodynamic interactions can be easily switched off in MPC algorithms, making it easy to study the importance of hydrodynamic interactions [24, 25].

It must, however, be emphasized that all local algorithms such as MPC, DPD, and LB model *compressible* fluids, so that it takes time for the hydrodynamic interactions to "propagate" over longer distances. As a consequence, these methods become quite inefficient in the Stokes limit, where the Reynolds number approaches zero. Algorithms which incorporate an Oseen tensor do not share this shortcoming.

The simplicity of the SRD algorithm has made it possible to derive analytic expressions for the transport coefficients which are valid for both large and small mean free paths [26–28]. This is usually very difficult to do for other mesoscale particle-based algorithms. Take DPD as an example: the viscosity measured in [29] is about 50% smaller than the value predicted theoretically in the same paper. For SRD, the agreement is generally better than 1%.

MPC is particularly well suited (1) for studying phenomena where both thermal fluctuations and hydrodynamics are important, (2) for systems with Reynolds and Peclet numbers of order 0.1–10, (3) if exact analytical expressions for the transport coefficients and consistent thermodynamics are needed, and (4) for modeling complex phenomena for which the constitutive relations are not known. Examples include chemically reacting flows, self-propelled objects, or solutions with embedded macromolecules and aggregates.

If thermal fluctuations are not essential or undesirable, a more traditional method such as a finite-element solver or a LB approach is recommended. If, on the other hand, inertia and fully resolved hydrodynamics are not crucial, but fluctuations are, one might be better served using Langevin or BD.

This review consists of two parts. The first part begins in Sect. 2 with a description of several widely used MPC algorithms and then discusses important features of the original SRD algorithm and a frequently used variation, Multi-Particle Collision Dynamics with Anderson Thermostat (MPC-AT), which effectively thermostats the system by replacing the relative velocities of particles in a collision cell with newly generated Gaussian random numbers in the collision step. After a qualitative discussion of the static and dynamic properties of MPC fluids in Sect. 3, two alternative approaches for deriving the hydrodynamic equations and evaluating the transport coefficients are described. First, in Sect. 4, discrete-time projection operator methods are discussed and the explicit form of the resulting Green–Kubo (GK) relations for the transport coefficients are given and evaluated. Subsequently, in Sect. 5, an alternative non-equilibrium approach is described. The two approaches complement each other, and the predictions of both methods are shown to be in complete agreement. It is then shown in Sect. 6 how MPC algorithms can be generalized to model non-ideal fluids and binary mixtures. Finally, various approaches for implementing slip and no-slip boundary conditions – as well as the coupling of embedded objects to a MPC solvent – are described in Sect. 7. In Sect. 8, the importance of angular-momentum conservation is discussed, in particular in systems of phase-separated fluids with different viscosities under flow. An important aspect of mesoscale simulations is the possibility to directly determine the effect of hydrodynamic interactions by switching them off, while retaining the same thermal fluctuations and similar friction coefficients; in MPC, this can be done very efficiently using an algorithm described in Sect. 9. The second part of the review describes a number of recent applications of MPC algorithms to study colloid and polymer dynamics, and the behavior of vesicles and cells in hydrodynamic flows. Section 10 focuses on the non-equilibrium behavior of colloidal suspensions, the dynamics of dilute solutions of linear polymers both in equilibrium and under flow conditions, and the properties of star polymers – also called ultra-soft colloids – in shear flow. Section 11 is devoted to the review of recent simulation results for vesicles in flow. After a short introduction to the modeling of membranes with different levels of coarse-graining, the behavior of fluid vesicles and red blood cells, both in shear and capillary flow, is discussed. Finally, a simple extension of MPC for viscoelastic solvents is described in Sect. 12, where the point particles of MPC for Newtonian fluids are replaced by harmonic dumbbells.

A discussion of several complementary applications – such as chemically reactive flows and self-propelled objects – can be found in a recent review of MPC by Kapral [30].

2 Algorithms

In the following, we use the term MPC to describe the generic class of particle-based algorithms for fluid flow which consist of successive free-streaming and multi-particle collision steps. The name SRD is reserved for the most widely used algorithm which was introduced by Malevanets and Kapral [18]. The name refers to the fact that the collisions consist of a *random rotation* of the relative velocities $\delta \mathbf{v}_i = \mathbf{v}_i - \mathbf{u}$ of the particles in a collision cell, where $\mathbf{u}$ is the mean velocity of all particles in a cell. There are a number of other MPC algorithms with different collision rules [31–33]. For example, one class of algorithms uses modified collision rules which provide a nontrivial "collisional" contribution to the equation of state [33, 34]. As a result, these models can be used to model non-ideal fluids or multi-component mixtures with a consolute point.

2.1   Stochastic Rotation Dynamics

In SRD, the solvent is modeled by a large number N of point-like particles of mass m which move in continuous space with a continuous distribution of velocities. The algorithm consists of individual streaming and collision steps. In the streaming step, the coordinates, $\mathbf{r}_i(t)$, of all solvent particles at time t are simultaneously updated according to

$$\mathbf{r}_i(t+\Delta t) = \mathbf{r}_i(t) + \Delta t\,\mathbf{v}_i(t)\,, \tag{1}$$

where $\mathbf{v}_i(t)$ is the velocity of particle i at time t and Δt is the value of the discretized time step.

In order to define the collisions, particles are sorted into cells, and they interact only with members of their own cell. Typically, the system is coarse-grained into cells of a regular, typically cubic, grid with lattice constant a. In practice, lengths are often measured in units of a, which corresponds to setting $a = 1$. The average number of particles per cell, M, is typically chosen to be between three and 20. The actual number of particles in a cell at a given time, which fluctuates, will be denoted by N_c. The collision step consists of a random rotation $\mathbf{R}$ of the relative velocities $\delta\mathbf{v}_i = \mathbf{v}_i - \mathbf{u}$ of all the particles in the collision cell,

$$\mathbf{v}_i(t+\Delta t) = \mathbf{u}(t) + \mathbf{R}\cdot\delta\mathbf{v}_i(t)\,. \tag{2}$$

All particles in the cell are subject to the same rotation, but the rotations in different cells and at different times are statistically independent. There is a great deal of freedom in how the rotation step is implemented, and any stochastic rotation matrix which satisfies semi-detailed balance can be used. Here, we describe the most commonly used algorithm. In two dimensions, $\mathbf{R}$ is a rotation by an angle $\pm\alpha$, with probability $1/2$. In three dimensions, a rotation by a fixed angle α about a randomly chosen axis is typically used. Note that rotations by an angle $-\alpha$ need not be considered, since this amounts to a rotation by an angle α about an axis with the opposite orientation. If we denote the randomly chosen rotation axis by $\hat{\mathbf{R}}$, the explicit collision rule in three dimensions is

$$\begin{aligned}\mathbf{v}_i(t+\Delta t) = \mathbf{u}(t) + \delta\mathbf{v}_{i,\perp}(t)\cos(\alpha)\\ + (\delta\mathbf{v}_{i,\perp}(t)\times\hat{\mathbf{R}})\sin(\alpha) + \delta\mathbf{v}_{i,\|}(t)\,,\end{aligned} \tag{3}$$

where $\perp$ and $\|$ are the components of the vector which are perpendicular and parallel to the random axis $\hat{\mathbf{R}}$, respectively. Malevanets and Kapral [18] have shown that there is an H-theorem for the algorithm, that the equilibrium distribution of velocities is Maxwellian, and that it yields the correct hydrodynamic equations with an ideal-gas equation of state.

In its original form [18,19], the SRD algorithm was not Galilean invariant. This is most pronounced at low temperatures or small time steps, where the mean free path, $\lambda = \Delta t\sqrt{k_{\mathrm{B}}T/m}$, is smaller than the cell size a. If the particles travel a distance between collisions which is small compared to the cell size, essentially the same

particles collide repeatedly before other particles enter the cell or some of the partic-
ipating particles leave the cell. For small λ, large numbers of particles in a given cell
remain correlated over several time steps. This leads to a breakdown of the molecular chaos assumption – i.e., particles become correlated and retain information of
previous encounters. Since these correlations are changed by a homogeneous imposed flow field, $\mathbf{V}$, Galilean invariance is destroyed, and the transport coefficients
depend on both the magnitude and direction of $\mathbf{V}$.

Ihle and Kroll [20, 21] showed that Galilean invariance can be restored by performing a random shift of the entire computational grid before every collision step.
The grid shift constantly groups particles into new collision neighborhoods; the
collision environment no longer depends on the magnitude of an imposed homogeneous flow field, and the resulting hydrodynamic equations are Galilean invariant
for arbitrary temperatures and Mach number. This procedure is implemented by
shifting all particles by the *same* random vector with components uniformly distributed in the interval $[-a/2, a/2]$ before the collision step. Particles are then shifted
back to their original positions after the collision.

In addition to restoring Galilean invariance, this grid-shift procedure accelerates momentum transfer between cells and leads to a collisional contribution to the
transport coefficients. If the mean free path λ is larger than $a/2$, the violation of
Galilean invariance without grid shift is negligible, and it is not necessary to use this
procedure.

2.1.1 SRD with Angular Momentum Conservation

As noted by Pooley and Yeomans [35] and confirmed in [28], the macroscopic stress
tensor of SRD is *not* symmetric in $\partial_\alpha v_\beta$. The reason for this is that the multi-
particle collisions do not, in general, conserve angular momentum. The problem
is particularly pronounced for small mean free paths, where asymmetric collisional
contributions to the stress tensor dominate the viscosity (see Sect. 4.1.1). In contrast,
for mean free paths larger than the cell size, where kinetic contributions dominate,
the effect is negligible.

An anisotropic stress tensor means that there is non-zero dissipation if the entire fluid undergoes a rigid-body rotation, which is clearly unphysical. However, as
emphasized in [28], this asymmetry is not a problem for most applications in the
incompressible (or small Mach number) limit, since the form of the Navier–Stokes
equation is not changed. This is in accordance with results obtained in SRD simulations of vortex shedding behind an obstacle [36], and vesicle [37] and polymer
dynamics [14]. In particular, it has been shown that the linearized hydrodynamic
modes are completely unaffected in two dimensions; in three dimensions only the
sound damping is slightly modified [28].

However, very recently Götze et al. [38] identified several situations involving
rotating flow fields in which this asymmetry leads to significant deviations from
the behavior of a Newtonian fluid. This includes (1) systems in which boundary
conditions are defined by torques rather than prescribed velocities, (2) mixtures of

liquids with a viscosity contrast, and (3) polymers with a locally high monomer density and a monomer–monomer distance on the order of or smaller than the lattice constant, a, embedded in a MPC fluid. A more detailed discussion will be presented in Sect. 8 below.

For the SRD algorithm, it is possible to restore angular momentum conservation by having the collision angle depend on the specific positions of the particles within a collision cell. Such a modification was first suggested by Ryder [39] for SRD in two dimensions. She showed that the angular momentum of the particles in a collision cell is conserved if the collision angle α is chosen such that

$$\sin(\alpha) = -2AB/(A^2 + B^2) \quad \text{and} \quad \cos(\alpha) = (A^2 - B^2)/(A^2 + B^2), \quad (4)$$

where

$$A = \sum_1^{N_c} [\mathbf{r}_i \times (\mathbf{v}_i - \mathbf{u})]|_z \quad \text{and} \quad B = \sum_1^{N_c} \mathbf{r}_i \cdot (\mathbf{v}_i - \mathbf{u}). \quad (5)$$

When the collision angles are determined in this way, the viscous stress tensor is symmetric. Note, however, that evaluating (4) is time-consuming, since the collision angle needs to be computed for every collision cell every time step. This typically increases the CPU time by a factor close to 2.

A general procedure for implementing angular-momentum conservation in multi-particle collision algorithms was introduced by Noguchi et al. [32]; it is discussed in the following section.

2.2 Multi-Particle Collision Dynamics with Anderson Thermostat

A stochastic rotation of the particle velocities relative to the center-of-mass velocity is not the only possibility for performing multi-particle collisions. In particular, MPC simulations can be performed directly in the canonical ensemble by employing an Anderson thermostat (AT) [31, 32]; the resulting algorithm will be referred to as MPC-AT$-a$. In this algorithm, instead of performing a rotation of the relative velocities, $\{\delta\mathbf{v}_i\}$, in the collision step, new relative velocities are generated. The components of $\{\delta\mathbf{v}_i^{\text{ran}}\}$ are Gaussian random numbers with variance $\sqrt{k_B T/m}$. The collision rule is [32, 38]

$$\mathbf{v}_i(t + \Delta t) = \mathbf{u}(t) + \delta\mathbf{v}_i^{\text{ran}} = \mathbf{u}(t) + \mathbf{v}_i^{\text{ran}} - \sum_{j \in \text{cell}} \mathbf{v}_j^{\text{ran}}/N_c, \quad (6)$$

where N_c is the number of particles in the collision cell, and the sum runs over all particles in the cell. It is important to note that MPC-AT is both a collision procedure and a thermostat. Simulations are performed in the canonical ensemble, and no additional velocity rescaling is required in non-equilibrium simulations, where there is viscous heating.

Just as SRD, this algorithm conserves momentum at the cell level but not angular momentum. Angular momentum conservation can be restored [32, 39] by imposing constraints on the new relative velocities. This leads to an angular-momentum conserving modification of MPC-AT [32, 38], denoted MPC-AT+a. The collision rule in this case is

$$\mathbf{v}_i(t + \Delta t) = \mathbf{u}(t) + \mathbf{v}_{i,\mathrm{ran}} - \sum_{\mathrm{cell}} \mathbf{v}_{i,\mathrm{ran}}/N_c$$

$$+ \left\{ m\Pi^{-1} \sum_{j \in \mathrm{cell}} \left[\mathbf{r}_{j,\mathrm{c}} \times (\mathbf{v}_j - \mathbf{v}_j^{\mathrm{ran}}) \right] \times \mathbf{r}_{i,\mathrm{c}} \right\}, \tag{7}$$

where Π is the moment of inertia tensor of the particles in the cell, and $\mathbf{r}_{i,\mathrm{c}} = \mathbf{r}_i - \mathbf{R}_\mathrm{c}$ is the relative position of particle i in the cell and $\mathbf{R}_\mathrm{c}$ is the center of mass of all particles in the cell.

When implementing this algorithm, an unbiased multi-particle collision is first performed, which typically leads to a small change of angular momentum, $\Delta \mathbf{L}$. By solving the linear equation $-\Delta \mathbf{L} = \Pi \cdot \omega$, the angular velocity ω which is needed to cancel the initial change of angular momentum is then determined. The last term in (7) restores this angular momentum deficiency. MPC-AT can be adapted for simulations in the micro-canonical ensemble by imposing an additional constraint on the values of the new random relative velocities [32].

2.2.1 Comparison of SRD and MPC-AT

Because d Gaussian random numbers per particle are required at every iteration, where d is the spatial dimension, the speed of the random number generator is the limiting factor for MPC-AT. In contrast, the efficiency of SRD is rather insensitive to the speed of the random number generator since only $d - 1$ uniformly distributed random numbers are needed in every box per iteration, and even a low quality random number generator is sufficient, because the dynamics is self-averaging. A comparison for two-dimensional systems shows that MPC-AT$-a$ is about a factor 2–3 times slower than SRD, and that MPC-AT+a is about a factor 1.3–1.5 slower than MPC-AT$-a$ [40].

One important difference between SRD and MPC-AT is the fact that relaxation times in MPC-AT generally *decrease* when the number of particles per cell is increased, while they *increase* for SRD. A longer relaxation time means that a larger number of time steps is required for transport coefficients to reach their asymptotic value. This could be of importance when fast oscillatory or transient processes are investigated. As a consequence, when using SRD, the average number of particles per cell should be in the range 3–20; otherwise, the internal relaxation times could be no longer negligible compared to physical time scales. No such limitation exists for MPC-AT, where the relaxation times scale as $(\ln M)^{-1}$, where M is the average number of particles in a collision cell.

2.3 Computationally Efficient Cell-Level Thermostating for SRD

The MPC-AT algorithm discussed in Sect. 2.2 provides a very efficient particle-level thermostating of the system. However, it is considerably slower than the original SRD algorithm, and there are situations in which the additional freedom offered by the choice of SRD collision angle can be useful.

Thermostating is required in any non-equilibrium MPC simulation, where there is viscous heating. A basic requirement of any thermostat is that it does not violate local momentum conservation, smear out local flow profiles, or distort the velocity distribution too much. When there is homogeneous heating, the simplest way to maintain a constant temperature is to just rescale velocity components by a scale factor S, $v_\alpha^{\text{new}} = S v_\alpha$, which adjusts the total kinetic energy to the desired value. This can be done with just a single global scale factor, or a local factor which is different in every cell. For a known macroscopic flow profile, $\mathbf{u}$, like in shear flow, the relative velocities $\mathbf{v} - \mathbf{u}$ can be rescaled. This is known as a profile-unbiased thermostat; however, it has been shown to have deficiencies in molecular dynamics simulations [41].

Here we describe an alternative thermostat which exactly conserves momentum in every cell and is easily incorporated into the MPC collision step. It was originally developed by Heyes for constant-temperature molecular dynamics simulations; however, the original algorithm described in [42] violates detailed balance. The thermostat consists of the following procedure which is performed independently in every collision cell as part of the collision step:

1. Randomly select a real number $\psi \in [1, 1+c]$, where c is a small number between 0.05 and 0.3 which determines the strength of the thermostat.
2. Accept this number as a scaling factor $S = \psi$ with probability $1/2$; otherwise, take $S = 1/\psi$.
3. Create another random number $\xi \in [0, 1]$. Rescale the velocities if ξ is smaller than the acceptance probability $p_A = \min(1, A)$, where

$$A = S^{d(N_c-1)} \exp\left[-\frac{m}{2 k_B T_0} \sum_{i=1}^{N_c} (\mathbf{v}_i - \mathbf{u})^2 \{S^2 - 1\} \right]. \tag{8}$$

   d is the spatial dimension, and N_c is the number of particles in the cell. The prefactor in (8) is an entropic contribution which accounts for the fact that the phase-space volume changes if the velocities are rescaled.
4. If the attempt is accepted, perform a stochastic rotation with the scaled rotation matrix $S\mathbf{R}$. Otherwise, use the rotation matrix $\mathbf{R}$.

This thermostat reproduces the Maxwell velocity distribution and does not change the viscosity of the fluid. It gives excellent equilibration, and the deviation of the measured kinetic temperature from T_0 is smaller than 0.01%. The parameter c controls the rate at which the kinetic temperature relaxes to T_0, and in agreement with experience from MC-simulations, an acceptance rate in the range of 50–65% leads to the fastest relaxation. For these acceptance rates, the relaxation time is of the order of 5–10 time steps. The corresponding value for c depends on the particle number N_c; in two dimensions, it is about 0.3 for $N_c = 7$ and decreases to 0.05 for $N_c = 100$. This thermostat has been successfully applied to SRD simulations of sedimenting charged colloids [16].

3 Qualitative Discussion of Static and Dynamic Properties

The previous section outlines several multi-particle algorithms. A detailed discussion of the link between the microscopic dynamics described by (1) and (2) or (3) and the macroscopic hydrodynamic equations, which describe the behavior at large length and time scales, requires a more careful analysis of the corresponding Liouville operator $\mathcal{L}$. Before describing this approach in more detail, we provide a more heuristic discussion of the equation of state and of one of the transport coefficients, the shear viscosity, using more familiar approaches for analyzing the behavior of dynamical systems.

3.1 Equation of State

In a homogeneous fluid, the pressure is the normal force exerted by the fluid on one side of a unit area on the fluid on the other side; expressed somewhat differently, it is the momentum transfer per unit area per unit time across an imaginary (flat) fixed surface. There are both *kinetic* and *virial* contributions to the pressure. The first arises from the momentum transported across the surface by particles that cross the surface in the unit time interval; it yields the ideal-gas contribution, $P_{id} = Nk_BT/V$, to the pressure. For classical particles interacting via pair-additive, central forces, the intermolecular "potential" contribution to the pressure can be determined using the method introduced by Irving and Kirkwood [43]. A clear discussion of this approach is given by Davis in [44], where it is shown to lead to the virial equation of state of a homogeneous fluid,

$$P = \frac{Nk_BT}{V} + \frac{1}{3V} \sum_i \langle \mathbf{r}_i \cdot \mathbf{F}_i \rangle, \tag{9}$$

in three dimensions, where $\mathbf{F}_i$ is the force on particle i due to all the other particles, and the sum runs over all particles of the system.

The kinetic contribution to the pressure, $P_{id} = Nk_BT/V$, is clearly present in all MPC algorithms. For SRD, this is the only contribution. The reason is that the stochastic rotations, which define the collisions, transport (on average) no net momentum across a fixed dividing surface. More general MPC algorithms (such as those discussed in Sect. 6) have an additional contribution to the virial equation of state. However, instead of an explicit force $\mathbf{F}_i$ as in (9), the contribution from the multi-particle collisions is a force of the form $m\Delta v_i/\Delta t$, and the role of the particle position, $\mathbf{r}_i$, is played by a variable which denotes the cell-partners which participate in the collision [33, 45].

3.2 Shear Viscosity

Just as for the pressure, there are both kinetic and collisional contributions to the transport coefficients. We present here a heuristic discussion of these contributions to the shear viscosity, since it illustrates rather clearly the essential physics and provides background for subsequent technical discussions.

Consider a reference plane (a line in two dimensions) normal in the y-direction embedded in a homogeneous fluid in equilibrium. The fluid below the plane exerts a mean force $\mathbf{p}_y$ per unit area on the fluid above the plane; by Newton's third law, the fluid above the plane must exert a mean force $-\mathbf{p}_y$ on the fluid below the plane. The normal force per unit area is just the mean pressure, P, so that $p_{yy} = P$. In a homogeneous simple fluid in which there are no velocity gradients, there is no tangential force, so that, for example, $p_{yx} = 0$. $p_{\alpha\beta}$ is called the *pressure tensor*, and the last result is just a statement of the well-known fact that the pressure tensor in a homogeneous simple fluid at equilibrium has no off-diagonal elements; the diagonal elements are all equal to the mean pressure P.

Consider a shear flow with a shear rate $\dot{\gamma} = \partial u_x(y)/\partial y$. In this case, there is a tangential stress on the reference surface because of the velocity gradient normal to the plane. In the small gradient limit, the *dynamic viscosity*, η, is defined as the coefficient of proportionality between the tangential stress, p_{yx}, and the normal gradient of the imposed velocity gradient,

$$p_{yx} = -\eta\dot{\gamma}. \tag{10}$$

The *kinematic viscosity*, ν, is related to η by $\nu = \eta/\rho$, where $\rho = nm$ is the mass density, with n the number density of the fluid and m the particle mass.

Kinetic contribution to the shear viscosity: The kinetic contribution to the shear viscosity comes from transverse momentum transport by the flow of fluid particles. This is the dominant contribution to the viscosity of gases. The following analogy may make this origin of viscosity clearer. Consider two ships moving side by side in parallel, but with different speeds. If the sailors on the two ships constantly throw sand bags from their ship onto the other, there will be a transfer of momentum between to two ships so that the slower ship accelerates and the faster ship decelerates. This can be interpreted as an effective friction, or kinetic viscosity, between ships. There are no direct forces between the ships, and the transverse momentum transfer originates solely from throwing sandbags from one ship to the other.

A standard result from kinetic theory is that the kinetic contribution to the shear viscosity in simple gases is [46]

$$\eta^{\text{kin}} \sim nm\bar{v}\lambda, \tag{11}$$

where λ is the mean free path and $\bar{v}$ is the thermal velocity. Using the fact that $\lambda \sim \bar{v}\Delta t$ for SRD and that $\bar{v} \sim \sqrt{k_B T/m}$, relation (11) implies that

$$\eta^{\text{kin}} \sim nk_B T \Delta t, \quad \text{or equivalently,} \quad v^{\text{kin}} \sim k_B T \Delta t/m, \tag{12}$$

which is, as more detailed calculations presented later will show, the correct dependence on n, $k_B T$, and Δt. In fact, the general form for the kinetic contribution to the kinematic viscosity is

$$v^{\text{kin}} = \frac{k_B T \Delta t}{m} f_{\text{kin}}(d, M, \alpha), \tag{13}$$

where d is the spatial dimension, M is the mean number of particles per cell, and α is the SRD collision angle. Another way of obtaining this result is to use the analogy with a random walk: The kinematic viscosity is the diffusion coefficient for momentum diffusion. At large mean free path, $\lambda/a \gg 1$, momentum is primarily transported by particle translation (as in the ship analogy). The mean distance a particle streams during one time step, Δt, is λ. According to the theory of random walks, the corresponding diffusion coefficient scales as $v^{\text{kin}} \sim \lambda^2/\Delta t \sim k_B T \Delta t/m$.

Note that in contrast to a "real" gas, for which the viscosity has a square root dependence on the temperature, $v^{\text{kin}} \sim T$ for SRD. This is because the mean free path of a particle in SRD does not depend on density; SRD allows particles to stream right through each other between collisions. Note, however, that SRD can be easily modified to give whatever temperature dependence is desired. For example, an additional temperature-dependent collision probability can be introduced; this would be of interest, e.g., for a simulation of realistic shock-wave profiles.

Collisional contribution to the shear viscosity: At small mean free paths, $\lambda/a \ll 1$, particles "stream" only a short distance between collisions, and the multi-particle "collisions" are the primary mechanism for momentum transport. These collisions redistribute momenta within cells of linear size a. This means that momentum "hops" an average distance a in one time step, leading to a momentum diffusion coefficient $v^{\text{col}} \sim a^2/\Delta t$. The general form of the collisional contribution to the shear viscosity is therefore

$$v^{\text{col}} = \frac{a^2}{\Delta t} f_{\text{col}}(d, M, \alpha). \tag{14}$$

This is indeed the scaling observed in numerical simulations at small mean free path.

The kinetic contribution dominates for $\lambda \gg a$, while the collisional contribution dominates in the opposite limit. Two other transport coefficients of interest are the thermal diffusivity, D_T, and the single particle diffusion coefficient, D. Both have the dimension square meter per second. As dimensional analysis would suggest, the kinetic and collisional contributions to D_T exhibit the same characteristic dependencies on λ, a, and Δt described by (13) and (14). Since there is no collisional contribution to the diffusion coefficient, $D \sim \lambda^2/\Delta t$.

Two complementary approaches have been used to derive the transport coefficients of the SRD fluid. The first is an equilibrium approach which utilizes a discrete projection operator formalism to obtain GK relations which express the transport coefficients as sums over the autocorrelation functions of reduced fluxes. This approach was first utilized by Malevanets and Kapral [19], and later extended by Ihle, Kroll and Tüzel [20,27,28] to include collisional contributions and arbitrary rotation angles. This approach is described in Sect. 4.1.

The other approach uses kinetic theory to calculate the transport coefficients in a stationary non-equilibrium situation such as shear flow. The first application of this approach to SRD was presented in [21], where the collisional contribution to the shear viscosity for large M, where particle number fluctuations can be ignored, was calculated. This scheme was later extended by Kikuchi et al. [26] to include fluctuations in the number of particles per cell, and then used to obtain expressions for the kinetic contributions to shear viscosity and thermal conductivity [35]. This non-equilibrium approach is described in Sect. 5.

4 Equilibrium Calculation of Dynamic Properties

A projection operator formalism for deriving the linearized hydrodynamic equations and GK relations for the transport coefficients of molecular fluids was originally introduced by Zwanzig [47–49] and later adapted for lattice gases by Dufty and Ernst [50]. With the help of this formalism, explicit expressions for both the reversible (Euler) as well as dissipative terms of the long-time, large-length-scale hydrodynamics equations for the coarse-grained hydrodynamic variables were derived. In addition, the resulting GK relations enable explicit calculations of the transport coefficients of the fluid. This work is summarized in Sect. 4.1. An analysis of the equilibrium fluctuations of the hydrodynamic modes can then be used to directly measure the shear and bulk viscosities as well as the thermal diffusivity. This approach is described in Sect. 4.2, where SRD results for the dynamic structure factor are discussed.

4.1 Linearized Hydrodynamics and Green–Kubo Relations

The GK relations for SRD differ from the well-known continuous versions due to the discrete-time dynamics, the underlying lattice structure, and the multi-particle interactions. In the following, we briefly outline this approach for determining the transport coefficients. More details can be found in [20, 27].

The starting point of this theory are microscopic definitions of local hydrodynamic densities A_β. These "slow" variables are the local number, momentum, and energy density. At the cell level, they are defined as

$$A_\beta(\xi) = \sum_{i=1}^{N} a_{\beta,i} \prod_{\gamma=1}^{d} \Theta\left(\frac{a}{2} - \left|\xi_\gamma + \frac{a}{2} - r_{i\gamma}\right|\right), \tag{15}$$

with the discrete cell coordinates $\xi = a\mathbf{m}$, where $m_\beta = 1,\ldots,L$, for each spatial component. $a_{1,i} = 1$ is the particle density, $\{a_{\beta,i}\} = m\{v_{i(\beta-1)}\}$, with $\beta = 2,\ldots,d+1$, are the components of the particle momenta, and $a_{d+2,i} = mv_i^2/2$ is the kinetic energy of particle i. d is the spatial dimension, and $\mathbf{r}_i$ and $\mathbf{v}_i$ are position and velocity of particle i, respectively.

$A_\beta(\xi)$, for $\beta = 2,\ldots,d+2$, are cell level coarse-grained densities. For example, $A_2(\xi)$ is the x-component of the total momentum of all the particles in cell ξ at the given time. Note that the particle density, A_1, was not coarse-grained in [20], i.e., the Θ functions in (15) were replaced by a δ-function. This was motivated by the fact that during collisions the particle number is trivially conserved in areas of arbitrary size, whereas energy and momentum are only conserved at the cell level.

The equilibrium correlation functions for the conserved variables are defined by $\langle\delta A_\beta(\mathbf{r},t)\delta A_\gamma(\mathbf{r}',t')\rangle$, where $\langle\delta A\rangle = A - \langle A\rangle$, and the brackets denote an average over the equilibrium distribution. In a stationary, translationally invariant system, the correlation functions depend only on the differences $\mathbf{r} - \mathbf{r}'$ and $t - t'$, and the Fourier transform of the matrix of correlation functions is

$$G_{\alpha\beta}(\mathbf{k},t) = \frac{1}{V}\langle\delta A_\beta^*(\mathbf{k},0)\delta A_\gamma(\mathbf{k},t)\rangle, \tag{16}$$

where the asterisk denotes the complex conjugate, and the spatial Fourier transforms of the densities are given by

$$A_\beta(\mathbf{k}) = \sum_j a_{\beta,j} e^{i\mathbf{k}\cdot\xi_j}, \tag{17}$$

where ξ_j is the coordinate of the cell occupied by particle j. $\mathbf{k} = 2\pi\mathbf{n}/(aL)$ is the wave vector, where $n_\beta = 0,\pm1,\ldots,\pm(L-1),L$ for the spatial components. To simplify notation, we omit the wave-vector dependence of $G_{\alpha\beta}$ in this section.

The collision invariants for the conserved densities are

$$\sum_j e^{i\mathbf{k}\cdot\xi_j^s(t+\Delta t)} \left[a_{\beta,j}(t+\Delta t) - a_{\beta,j}(t)\right] = 0, \tag{18}$$

where ξ_j^s is the coordinate of the cell occupied by particle j in the *shifted* system. Starting from these conservation laws, a projection operator can be constructed that projects the full SRD dynamics onto the conserved fields [20]. The central result is that the discrete Laplace transform of the linearized hydrodynamic equations can be written as

$$\left[s + ik\Omega + k^2\Lambda\right] G(\mathbf{k},s) = \frac{1}{\Delta t} G(0)R(k), \tag{19}$$

where $R(k) = [1 + \Delta t(ik\Omega + k^2\Lambda)]^{-1}$ is the residue of the hydrodynamic pole [20]. The linearized hydrodynamic equations describe the long-time large-length-scale dynamics of the system, and are valid in the limits of small k and s. The frequency matrix Ω contains the reversible (Euler) terms of the hydrodynamic equations. Λ is the matrix of transport coefficients. The discrete GK relation for the matrix of viscous transport coefficients is [20]

$$\Lambda_{\alpha\beta}(\hat{\mathbf{k}}) \equiv \frac{\Delta t}{Nk_{\mathrm{B}}T} \sum_{t=0}^{\infty}{}' \langle \hat{k}_\lambda \sigma_{\alpha\lambda}(0) | \hat{k}_{\lambda'} \sigma_{\beta\lambda'}(t) \rangle, \tag{20}$$

where the prime on the sum indicates that the $t = 0$ term has the relative weight $1/2$. $\sigma_{\alpha\beta} = P\delta_{\alpha\beta} - p_{\alpha\beta}$ is the viscous stress tensor. The reduced fluxes in (20) are given by

$$\hat{k}_\lambda \sigma_{\alpha\lambda}(t) = \frac{m}{\Delta t} \sum_j \left(-v_{j\alpha}(t)\hat{\mathbf{k}} \cdot \left[\Delta \xi_j(t) + \Delta v_{j\alpha}(t) \Delta \xi_j^s(t) \right] + \frac{\Delta t}{d} \hat{k}_\alpha v_j^2(t) \right) \tag{21}$$

for $\alpha = 1,\ldots,d$, with $\Delta\xi_j(t) = \xi_j(t + \Delta t) - \xi_j(t)$, $\Delta\xi_j^s(t + \Delta t) = \xi_j(t + \Delta t) - \xi_j^s(t + \Delta t)$, and $\Delta v_{xj}(t) = v_{xj}(t + \Delta t) - v_{xj}(t)$. $\xi_j(t)$ is the cell coordinate of particle j at time t, while ξ_j^s is its cell coordinate in the (stochastically) shifted frame. The corresponding expressions for the thermal diffusivity and self-diffusion coefficient can be found in [20].

The straightforward evaluation of the GK relations for the viscous (21) and thermal transport coefficients leads to three – kinetic, collisional, and mixed – contributions. In addition, it was found that for mean free paths λ smaller than the cell size a, there are finite cell-size corrections which could not be summed in a controlled fashion. The origin of the problem was the explicit appearance of $\Delta\xi$ in the stress correlations. However, it was subsequently shown [28, 51] that the GK relations can be re-summed by introducing a stochastic variable, B_i, which is the difference between change in the shifted cell coordinates of particle i during one streaming step and the actual distance traveled, $\Delta t\, \mathbf{v}_i$. The resulting microscopic stress tensor for the viscous modes is

$$\bar{\sigma}_{\alpha\beta} = \sum_i \left[mv_{i\alpha}v_{i\beta} + \frac{m}{\Delta t} v_{i\alpha} B_{i\beta} \right] \tag{22}$$

where $B_{j\beta}(t) = \xi_{j\beta}^s(t + \Delta t) - \xi_{j\beta}^s(t) - \Delta t\, v_{j\beta}(t)$. It is interesting to compare this result to the corresponding expression

$$\sigma_{\alpha\beta} = \sum_i \delta(\mathbf{r} - \mathbf{r}_i) \left[mv_{i\alpha}v_{i\beta} + \frac{1}{2} \sum_{j\neq i} r_{ij\alpha} \mathcal{F}_{ij\beta}(\mathbf{r}_{ij}) \right] \tag{23}$$

for molecular fluids. The first term in both expressions, the ideal-gas contribution, is the same in both cases. The collisional contributions, however, are quite different.

The primary reason is that in SRD, the collisional contribution corresponds to a non-local (on the scale of the cell size) force which acts only at discrete time intervals.

B_i has a number of important properties which simplify the calculation of the transport coefficients. In particular, it is shown in [28,51] that stress–stress correlation functions involving one B_i in the GK relations for the transport coefficients are zero, so that, for example, $\Lambda_{\alpha\beta}(\hat{\mathbf{k}}) = \Lambda_{\alpha\beta}^{\mathrm{kin}}(\hat{\mathbf{k}}) + \Lambda_{\alpha\beta}^{\mathrm{col}}(\hat{\mathbf{k}})$, with

$$\Lambda_{\alpha\beta}^{\mathrm{kin}}(\hat{\mathbf{k}}) = \frac{\Delta t}{Nmk_{\mathrm{B}}T} \sum_{n=0}^{\infty}{}' \langle \hat{k}_\lambda \sigma_{\alpha\lambda}^{\mathrm{kin}}(0) | \hat{k}_{\lambda'} \sigma_{\beta\lambda}^{\mathrm{kin}}(n\Delta t) \rangle \tag{24}$$

and

$$\Lambda_{\alpha\beta}^{\mathrm{col}}(\hat{\mathbf{k}}) = \frac{\Delta t}{Nmk_{\mathrm{B}}T} \sum_{n=0}^{\infty}{}' \langle \hat{k}_\lambda \sigma_{\alpha\lambda}^{\mathrm{col}}(0) | \hat{k}_{\lambda'} \sigma_{\beta\lambda}^{\mathrm{col}}(n\Delta t) \rangle], \tag{25}$$

with

$$\sigma_{\alpha\beta}^{\mathrm{kin}}(n\Delta t) = \sum_j mv_{j\alpha}(n\Delta t)v_{j\beta}(n\Delta t) \tag{26}$$

and

$$\sigma_{\alpha\beta}^{\mathrm{col}}(n\Delta t) = \frac{1}{\Delta t} \sum_j mv_{j\alpha}(n\Delta t)B_{j\beta}(n\Delta t), \tag{27}$$

where $B_{j\beta}(n\Delta t) = \xi_{j\beta}^s([n+1]\Delta t) - \xi_{j\beta}^s(n\Delta t) - \Delta t v_{j\beta}(n\Delta t)$. Similar relations were obtained for the thermal diffusivity in [28].

4.1.1 Explicit Expressions for the Transport Coefficients

Analytical calculations of the SRD transport coefficients are greatly simplified by the fact that collisional and kinetic contributions to the stress–stress autocorrelation functions decouple. Both the kinetic and collisional contributions have been calculated explicitly in two and three dimension, and numerous numerical tests have shown that the resulting expressions for all the transport coefficients are in excellent agreement with simulation data. Before summarizing the results of this work, it is important to emphasize that because of the cell structure introduced to define coarse-grained collisions, angular momentum is not conserved in a collision [28,35,39]. As a consequence, the macroscopic viscous stress tensor is not, in general, a symmetric function of the derivatives $\partial_\alpha v_\beta$. Although the kinetic contributions to the transport coefficients lead to a symmetric stress tensor, the collisional do not. Before evaluating the transport coefficients, we discuss the general form of the macroscopic viscous stress tensor.

Assuming only cubic symmetry and allowing for a non-symmetric stress tensor, the most general form of the linearized Navier–Stokes equation is

$$\partial_t v_\alpha(\mathbf{k}) = -\partial_\alpha p + \Lambda_{\alpha\beta}(\hat{\mathbf{k}})v_\beta(\mathbf{k}), \tag{28}$$

where

$$\Lambda_{\alpha\beta}(\hat{\mathbf{k}}) \equiv v_1 \left(\delta_{\alpha,\beta} + \frac{d-2}{d} \hat{k}_\alpha \hat{k}_\beta \right) \tag{29}$$

$$+ v_2 \left(\delta_{\alpha,\beta} - \hat{k}_\alpha \hat{k}_\beta \right) + \gamma \hat{k}_\alpha \hat{k}_\beta + \kappa \, \hat{k}_\alpha^2 \delta_{\alpha,\beta}.$$

In a normal simple liquid, $\kappa = 0$ (because of invariance with respect to infinitesimal rotations) and $v_2 = 0$ (because the stress tensor is symmetric in $\partial_\alpha v_\beta$), so that the kinematic shear viscosity is $v = v_1$. In this case, (29) reduces to the well-known form [20]

$$\Lambda_{\alpha\beta}(\hat{\mathbf{k}}) = v \left(\delta_{\alpha,\beta} + \frac{d-2}{d} \hat{k}_\alpha \hat{k}_\beta \right) + \gamma \hat{k}_\alpha \hat{k}_\beta, \tag{30}$$

where γ is the bulk viscosity.

Kinetic contributions: Kinetic contributions to the transport coefficients dominate when the mean free path is larger than the cell size, i.e., $\lambda > a$. As can be seen from (24) and (26), an analytic calculation of these contributions requires the evaluation of time correlation functions of products of the particle velocities. This is straightforward if one makes the basic assumption of *molecular chaos* that successive collisions between particles are not correlated. In this case, the resulting time-series in (24) is geometrical, and can be summed analytically. The resulting expression for the shear viscosity in two dimensions is

$$v^{\text{kin}} = \frac{k_{\text{B}} T \Delta t}{2m} \left[\frac{M}{(M-1+e^{-M}) \sin^2(\alpha)} - 1 \right]. \tag{31}$$

Fluctuations in the number of particles per cell are included in (31). This result agrees with the non-equilibrium calculations of Pooley and Yeomans [35, 52], measurements in shear flow [26], and the numerical evaluation of the GK relation in equilibrium simulations (see Fig. 1).

The corresponding result in three dimensions for collision rule (3) is

$$v^{\text{kin}} = \frac{k_{\text{B}} T \Delta t}{2m} \left\{ \frac{5M}{(M-1+e^{-M})[2-\cos(\alpha)-\cos(2\alpha)]} - 1 \right\}. \tag{32}$$

The kinetic contribution to the stress tensor is symmetric, so that $v_2^{\text{kin}} = 0$ and the kinetic contribution to the shear viscosity is $v^{\text{kin}} \equiv v_1^{\text{kin}}$.

Collisional contributions: Explicit expressions for the collisional contributions to the viscous transport coefficients can be obtained by considering various choices for $\hat{\mathbf{k}}$ and α and β in (25), (27), and (29). Taking $\hat{\mathbf{k}}$ in the y-direction and $\alpha = \beta = 1$ yields

$$v_1^{\text{col}} + v_2^{\text{col}} = \frac{1}{\Delta t N k_{\text{B}} T} \sum_{t=0}^{\infty} {}' \sum_{i,j} \langle v_{ix}(0) B_{iy}(0) v_{ix}(t) B_{iy}(t) \rangle. \tag{33}$$

Other choices lead to relations between the collisional contributions to the viscous transport coefficients, namely

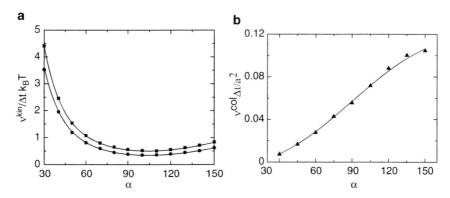

Fig. 1 **a** Normalized kinetic contribution to the viscosity, $v^{\text{kin}}/(\Delta t k_B T)$, in three dimensions as a function of the collision angle α. Data were obtained by time averaging the GK relation over 75,000 iterations using $\lambda/a = 2.309$ for $M = 5$ (*filled squares*) and $M = 20$ (*filled circles*). The *lines* are the theoretical prediction, (32). Parameters: $L/a = 32$, $\Delta t = 1$. From [53]. **b** Normalized collisional contribution to the viscosity, $v^{\text{col}} \Delta t/a^2$, in three dimensions as a function of the collision angle α. The *solid line* is the theoretical prediction, (39). Data were obtained by time averaging the GK relation over 300,000 iterations. Parameters: $L/a = 16$, $\lambda/a = 0.1$, $M = 3$, and $\Delta t = 1$. From [54]

$$[1 + (d-2)/d] \, v_1^{\text{col}} + \gamma^{\text{col}} + \kappa^{\text{col}} = v_1^{\text{col}} + v_2^{\text{col}}. \tag{34}$$

and

$$[(d-2)/d] \, v_1^{\text{col}} - v_2^{\text{col}} + \gamma^{\text{col}} = 0. \tag{35}$$

These results imply that $\kappa^{\text{col}} = 0$, and $\gamma^{\text{col}} - 2v_1^{\text{col}}/d = v_2^{\text{col}} - v_1^{\text{col}}$. It follows that the collision contribution to the macroscopic viscous stress tensor is

$$\begin{aligned}
\hat{\sigma}_{\alpha\beta}^{\text{col}}/\rho &= v_1^{\text{col}}(\partial_\beta v_\alpha + \partial_\alpha v_\beta) + v_2^{\text{col}}(\partial_\beta v_\alpha - \partial_\alpha v_\beta) + (v_2^{\text{col}} - v_1^{\text{col}})\delta_{\alpha\beta}\partial_\lambda v_\lambda \\
&= (v_1^{\text{col}} + v_2^{\text{col}})\partial_\beta v_\alpha + (v_2^{\text{col}} - v_1^{\text{col}})Q_{\alpha\beta},
\end{aligned} \tag{36}$$

where $Q_{\alpha\beta} \equiv \delta_{\alpha\beta}\partial_\lambda v_\lambda - \partial_\alpha v_\beta$. Since $Q_{\alpha\beta}$ has zero divergence, $\partial_\beta Q_{\alpha\beta} = 0$, the term containing Q in (36) will not appear in the linearized hydrodynamic equation for the momentum density, so that

$$\rho \frac{\partial \mathbf{v}}{\partial t} = -\nabla p + \rho(v^{\text{kin}} + v^{\text{col}})\Delta\mathbf{v} + \frac{d-2}{d}v^{\text{kin}}\nabla(\nabla \cdot \mathbf{v}), \tag{37}$$

where $v^{\text{col}} = v_1^{\text{col}} + v_2^{\text{col}}$. In writing (37) we have used the fact that the kinetic contribution to the microscopic stress tensor, $\bar{\sigma}^{\text{kin}}$, is symmetric, and $\gamma^{\text{kin}} = 0$ [27]. The viscous contribution to the sound attenuation coefficient is $v^{\text{col}} + 2(d-1)v^{\text{kin}}/d$ instead of the standard result, $2(d-1)v/d + \gamma$, for simple isotropic fluids. The collisional contribution to the effective shear viscosity is $v^{\text{col}} \equiv v_1^{\text{col}} + v_2^{\text{col}}$. It is interesting to note that the kinetic theory approach discussed in [35] is able to show explicitly that $v_1^{col} = v_2^{col}$, so that $v^{col} = 2v_1^{col}$.

It is straightforward to evaluate the various contributions to the right-hand side of (33). In particular, note that since velocity correlation functions are only required at equal times and for a time lag of one time step, molecular chaos can be assumed [51]. Using the relation [28]

$$\langle B_{i\alpha}(n\Delta t) B_{j\beta}(m\Delta t) \rangle = \frac{a^2}{12} \delta_{\alpha\beta} (1 + \delta_{ij}) [2\delta_{n,m} - \delta_{n,m+1} - \delta_{n,m-1}] , \qquad (38)$$

and averaging over the number of particles in a cell assuming that the number of particles in any cell is Poisson distributed at each time step, with an average number M of particles per cell, one then finds

$$v^{col} = v_1^{col} + v_2^{col} = \frac{a^2}{6d\Delta t} \left(\frac{M - 1 + e^{-M}}{M} \right) [1 - \cos(\alpha)] , \qquad (39)$$

for the SRD collision rules in both two and three dimensions. Equation (39) agrees with the result of [26] and [35] obtained using a completely different non-equilibrium approach in shear flow. Simulation results for the collisional contribution to the viscosity are in excellent agreement with this result (see Fig. 1).

Thermal diffusivity and self-diffusion coefficient: As with the viscosity, there are both kinetic and collisional contributions to the thermal diffusivity, D_T. A detailed analysis of both contributions is given in [28], and the results are summarized in Table 1. The self-diffusion coefficient, D, of particle i is defined by

$$D = \lim_{t \to \infty} \frac{1}{2dt} \langle [\mathbf{r}_i(t) - \mathbf{r}_i(0)]^2 \rangle = \frac{\Delta t}{d} \sum_{n=0}^{\infty} {}' \langle \mathbf{v}_i(n\Delta t) \cdot \mathbf{v}_i(0) \rangle, \qquad (40)$$

where the second expression is the corresponding discrete GK relation. The self-diffusion coefficient is unique in that the collisions do not explicitly contribute to D. With the assumption of molecular chaos, the kinetic contributions are easily summed [27] to obtain the result given in Table 1.

4.1.2 Beyond Molecular Chaos

The kinetic contributions to the transport coefficients presented in Table 1 have all been derived under the assumption of molecular chaos, i.e., that particle velocities are not correlated. Simulation results for the shear viscosity and thermal diffusivity have generally been found to be in good agreement with these results. However, it is known that there are correlation effects for λ/a smaller than unity [15, 55]. They arise from correlated collisions between particles that are in the same collision cell for more than one time step.

Table 1 Theoretical expressions for the kinematic shear viscosity ν, the thermal diffusivity, D_T, and the self-diffusion coefficient, D, in both two ($d = 2$) and three ($d = 3$) dimensions. M is the average number of particles per cell, α is the collision angle, k_B is Boltzmann's constant, T is the temperature, Δt is the time step, m is the particle mass, and a is the cell size. Except for self-diffusion constant, for which there is no collisional contribution, both the kinetic and collisional contributions are listed. The expressions for shear viscosity and self-diffusion coefficient include the effect of fluctuations in the number of particles per cell; however, for brevity, the relations for thermal diffusivity are correct only up to $O(1/M)$ and $O(1/M^2)$ for the kinetic and collisional contributions, respectively. For the complete expressions, see [28, 53, 54]

	d	Kinetic $(\times k_B T \Delta t / 2m)$	Collisional $(\times a^2/\Delta t)$
ν	2	$\dfrac{M}{(M-1+e^{-M})\sin^2(\alpha)} - 1$	$\dfrac{(M-1+e^{-M})}{6dM}[1-\cos(\alpha)]$
	3	$\dfrac{5M}{(M-1+e^{-M})[2-\cos(\alpha)-\cos(2\alpha)]} - 1$	
D_T	2	$\dfrac{d}{1-\cos(\alpha)} - 1 + \dfrac{2d}{M}\left[\dfrac{7-d}{5} - \dfrac{1}{4}\csc^2(\alpha/2)\right]$	$\dfrac{(1-1/M)}{3(d+2)M}[1-\cos(\alpha)]$
	3		
D	2	$\dfrac{dM}{[1-\cos(\alpha)](M-1+e^{-M})} - 1$	$-$
	3		

For the viscosity and thermal conductivity, these corrections are generally negligible, since they are only significant in the small λ/a regime, where the collisional contribution to the transport coefficients dominates. In this regard, it is important to note that there are no correlation corrections to ν^{col} and D_T^{col} [28]. For the self-diffusion coefficient – for which there is no collisional contribution – correlation corrections dramatically increase the value of this transport coefficient for $\lambda \ll a$, see [15, 55]. These correlation corrections, which arise from particles which collide with the same particles in consecutive time steps, are distinct from the correlation effects which are responsible for the long-time tails. This distinction is important, since long-time tails are also visible at large mean free paths, where these corrections are negligible.

4.2 Dynamic Structure Factor

Spontaneous thermal fluctuations of the density, $\rho(\mathbf{r}, t)$, the momentum density, $\mathbf{g}(\mathbf{r}, t)$, and the energy density, $\epsilon(\mathbf{r}, t)$, are dynamically coupled, and an analysis of their dynamic correlations in the limit of small wave numbers and frequencies can be used to measure a fluid's transport coefficients. In particular, because it is easily measured in dynamic light scattering, X-ray, and neutron scattering experiments, the Fourier transform of the density-density correlation function – the dynamics structure factor – is one of the most widely used vehicles for probing the dynamic and transport properties of liquids [56].

A detailed analysis of equilibrium dynamic correlation functions – the dynamic structure factor as well as the vorticity and entropy-density correlation functions – using the SRD algorithm is presented in [57]. The results – which are in good agreement with earlier numerical measurements and theoretical predictions – provided further evidence that the analytic expressions or the transport coefficients are accurate and that we have an excellent understanding of the SRD algorithm at the kinetic level.

Here, we briefly summarize the results for the dynamic structure factor. The dynamic structure factor exhibits three peaks, a central "Rayleigh" peak caused by the thermal diffusion, and two symmetrically placed "Brillouin peaks" caused by sound. The width of the central peak is determined by the thermal diffusivity, D_T, while that of the two Brillouin peaks is related to the sound attenuation coefficient, Γ. For the SRD algorithm [57],

$$\Gamma = D_T \left(\frac{c_p}{c_v} - 1 \right) + 2 \left(\frac{d-1}{d} \right) v^{\text{kin}} + v^{\text{col}}. \tag{41}$$

Note that in two-dimensions, the sound attenuation coefficient for a SRD fluid has the same functional dependence on D_T and $v = v^{\text{kin}} + v^{\text{col}}$ as an isotropic fluid with an ideal-gas equation of state (for which $\gamma = 0$).

Simulation results for the structure factor in two-dimensions with $\lambda/a = 1.0$ and collision angle $\alpha = 120°$, and $\lambda/a = 0.1$ with collision angle $\alpha = 60°$ are shown in Figs. 2a and 2b, respectively. The solid lines are the theoretical prediction for the dynamic structure factor (see (36) of [57]) using $c = \sqrt{2k_B T/m}$ and values for the transport coefficients obtained using the expressions in Table 1, assuming that the bulk viscosity $\gamma = 0$. As can be seen, the agreement is excellent.

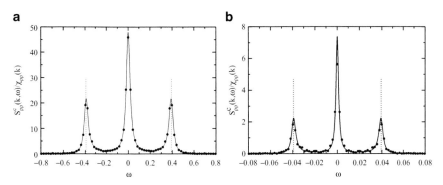

Fig. 2 Normalized dynamic structure, $S^c_{\rho\rho}(k\omega)/\chi_{\rho\rho}(k)$, for $\mathbf{k} = 2\pi(1,1)/L$ and **a** $\lambda/a = 1.0$ with $\alpha = 120°$, and **b** $\lambda/a = 0.1$ with $\alpha = 60°$. The *solid lines* are the theoretical prediction for the dynamic structure factor (see (36) of [57]) using values for the transport coefficients obtained with the expressions in Table 1. The *dotted lines* show the predicted positions of the Brillouin peaks, $\omega = \pm ck$, with $c = \sqrt{2k_B T/m}$. Parameters: $L/a = 32$, $M = 15$, and $\Delta t = 1.0$. From [57]

5   Non-Equilibrium Calculations of Transport Coefficients

MPC transport coefficients have also been evaluated by calculating the linear response of the system to imposed gradients. This approach was introduced by
Kikuchi et al. [26] for the shear viscosity and then extended and refined in [35]
to determine the thermal diffusivity and bulk viscosity. Here, we summarize the
derivation of the shear viscosity.

5.1   Shear Viscosity of SRD: Kinetic Contribution

Linear response theory provides an alternative, and complementary, approach for
evaluating the shear viscosity. This non-equilibrium approach is related to equilibrium calculations described in the previous section through the fluctuation–
dissipation theorem. Both methods yield identical results. For the more complicated
analysis of the hydrodynamic equations, the stress tensor, and the longitudinal transport coefficients such as the thermal conductivity, the reader is referred to [35].

Following Kikuchi et al. [26], we consider a two-dimensional liquid with an im-
posed shear $\dot{\gamma} = \partial u_x(y)/\partial y$. On average, the velocity profile is given by $\mathbf{v} = (\dot{\gamma}y, 0)$.
The dynamic shear viscosity η is the proportionality constant between the velocity
gradient $\dot{\gamma}$ and the frictional force acting on a plane perpendicular to y; i.e.,

$$\sigma_{xy} = \eta\dot{\gamma}, \tag{42}$$

where σ_{xy} is the off-diagonal element of the viscous stress tensor. During the stream-
ing step, particles will cross this plane only if $|v_y\Delta t|$ is greater than the distance to
the plane. Assuming that the fluid particles are homogeneously distributed, the momentum flux is obtained by integrating over the coordinates and velocities of all
particles that cross the plane from above and below during the time step Δt. The
result is [26]

$$\sigma_{xy} = \rho \left(\frac{\dot{\gamma}\Delta t}{2} \langle v_y^2 \rangle - \langle v_x v_y \rangle \right), \tag{43}$$

where the mass density $\rho = mM/a^d$, and the averages are taken over the steady-state
distribution $P(v_x - \dot{\gamma}y, v_y)$. It is important to note that this is *not* the Maxwell–
Boltzmann distribution, since we are in a non-equilibrium steady state where the
shear has induced correlations between v_x and v_y. As a consequence, $\langle v_x v_y \rangle$ is
nonzero. To determine the behavior of $\langle v_x v_y \rangle$, the effect of streaming and collisions are calculated separately. During streaming, particles which arrive at y_0
with positive velocity v_y have started from $y_0 - v_y\Delta t$; these particles bring a velocity component v_x which is smaller than that of particles originally located at
y_0. On the other hand, particles starting out at $y > y_0$ with negative v_y bring a
larger v_x. The velocity distribution is therefore sheared by the streaming, so that

$P^{\text{after}}(v_x, v_y) = P^{\text{before}}(v_x + \dot{\gamma} v_y \Delta t, v_y)$. Averaging $v_x v_y$ over this distribution gives [26]

$$\langle v_x v_y \rangle^{\text{after}} = \langle v_x v_y \rangle - \dot{\gamma} \Delta t \langle v_y^2 \rangle, \tag{44}$$

where the superscript denotes the quantity *after* streaming. The streaming step therefore reduces correlations by $-\dot{\gamma} \Delta t \langle v_y^2 \rangle$, making v_x and v_y increasingly anti-correlated.

The collision step redistributes momentum between particles and tends to reduce correlations. Making the assumption of molecular chaos, i.e., that the velocities of different particles are uncorrelated, and averaging over the two possible rotation directions, one finds

$$\langle v_x v_y \rangle^{\text{after}} = \left[1 - \frac{N_c - 1}{N_c} [1 - \cos(2\alpha)] \right] \langle v_x v_y \rangle^{\text{before}}. \tag{45}$$

The number of particles in a cell, N_c is not constant, and density fluctuations have to be included. The probability to find n uncorrelated particles in a given cell is given by the Poisson distribution, $w(n) = \exp(-M) M^n / n!$; the probability of a given particle being in a cell together with $n - 1$ others is $nw(n)/M$. Taking an average over this distribution gives

$$\langle v_x v_y \rangle^{\text{after}} = f \langle v_x v_y \rangle^{\text{before}}, \tag{46}$$

with

$$f = \left\{ 1 - \frac{M - 1 + \exp(-M)}{M} [1 - \cos(2\alpha)] \right\}. \tag{47}$$

The difference between this result and just replacing N_c by M in (45) is small, and only important for $M \leq 3$. One sees that $\langle v_x v_y \rangle$ is first modified by streaming and then multiplied by a factor f in the subsequent collision step. In the steady state, it therefore oscillates between two values. Using (44), (46), and (47), we obtain the self-consistency condition $(\langle v_x v_y \rangle - \dot{\gamma} \Delta t \langle v_y^2 \rangle) f = \langle v_x v_y \rangle$. Solving for $\langle v_x v_y \rangle$, assuming equipartition of energy, $\langle v_y^2 \rangle = k_B T / m$, and substituting into (43), we have

$$\sigma_{xy} = \frac{\dot{\gamma} M \Delta t k_B T}{m} \left(\frac{1}{2} + \frac{f}{1 - f} \right), \tag{48}$$

Inserting this result into the definition of the viscosity, (42), yields the same expression for the kinetic viscosity in two-dimensions as obtained by the equilibrium GK approach discussed in Sect. 4.1.1.

5.2 Shear Viscosity of SRD: Collisional Contribution

The collisional contribution to the shear viscosity is proportional to $a^2 / \Delta t$; as discussed in Sect. 3.2, it results from the momentum transfer between particles in a cell of size a during the collision step. Consider again a collision cell of linear dimension

a with a shear flow $u_x(y) = \dot{\gamma} y$. Since the collisions occur in a shifted grid, they cause a transfer of momentum between neighboring cells of the original unshifted reference frame [21, 27]. Consider now the momentum transfer due to collisions across the line $y = h$, the coordinate of a cell boundary in the unshifted frame. If we assume a homogeneous distribution of particles in the collision cell, the mean velocities in the upper $(y > h)$ and lower partitions are

$$\mathbf{u}_1 = \frac{1}{M_1} \sum_{i=1}^{M_1} \mathbf{v}_i \quad \text{and} \quad \mathbf{u}_2 = \frac{1}{M_2} \sum_{i=M_1+1}^{M} \mathbf{v}_i, \tag{49}$$

respectively, where $M_1 = M(a-h)/a$ and $M_2 = Mh/a$. Collisions transfer momentum between the two parts of the cell. The x-component of the momentum transfer is

$$\Delta p_x(h) \equiv \sum_{i=1}^{M_1} \left[v_{ix}^{\text{after}} - v_{ix}^{\text{before}} \right]. \tag{50}$$

The use of the rotation rule (2) together with an average over the sign of the stochastic rotation angle yields

$$\Delta p_x(h) = [\cos(\alpha) - 1] M_1 (u_{1x} - u_x). \tag{51}$$

Since $M\mathbf{u} = M_1 \mathbf{u}_1 + M_2 \mathbf{u}_2$,

$$\Delta p_x(h) = [1 - \cos(\alpha)] M (u_{2x} - u_{1x}) \frac{h}{a} \left(1 - \frac{h}{a}\right). \tag{52}$$

Averaging over the position h of the dividing line, which corresponds to averaging over the random shift, we find

$$\langle \Delta p_x \rangle = \frac{1}{a} \int_0^a \Delta p_x(h) dh = \frac{1}{6} [1 - \cos(\alpha)] M (u_{2x} - u_{1x}). \tag{53}$$

Since the dynamic viscosity η is defined as the ratio of the tangential stress, σ_{yx}, to $\partial u_x/\partial y$, we have

$$\eta = \frac{\langle \Delta p_x \rangle/(a^2 \Delta t)}{\partial u_x/\partial y} = \frac{\langle \Delta p_x \rangle/(a^2 \Delta t)}{(u_{2x} - u_{1x})/(a/2)}, \tag{54}$$

so that the kinematic viscosity, $\nu = \eta/\rho$, in two-dimensions for SRD is

$$\nu^{\text{col}} = \frac{a^2}{12\Delta t} [1 - \cos(\alpha)] \tag{55}$$

in the limit of small mean free path. Since we have neglected the fluctuations in the particle number, this expression corresponds to the limit $M \to \infty$. Even though this derivation is somewhat heuristic, it gives a remarkably accurate expression; in particular, it contains the correct dependence on the cell size, a, and the time step, Δt, in the limit of small free path,

$$v^{\text{col}} = \frac{a^2}{\Delta t} f_{\text{col}}(d, M, \alpha), \tag{56}$$

as expected from simple random walk arguments. Kikuchi et al. [26] included particle number fluctuations and obtained identical results for the collisional contribution to the viscosity as was obtained in the GK approach (see Table 1).

5.3 Shear Viscosity of MPC-AT

For MPC-AT, the viscosities have been calculated in [32] using the methods described in Sects. 5.1 and 5.2. The total viscosity of MPC-AT is given by the sum of two terms, the collisional and kinetic contributions. For MPC-AT$-a$, it was found for both two and three dimensions that [32]

$$v^{\text{kin}} = \frac{k_B T \Delta t}{m} \left(\frac{M}{M - 1 + e^{-M}} - \frac{1}{2} \right) \quad \text{and}$$

$$v^{\text{col}} = \frac{a^2}{12 \Delta t} \left(\frac{M - 1 + e^{-M}}{M} \right). \tag{57}$$

The exponential terms e^{-M} are due to the fluctuation of the particle number per cell and become important for $M \leq 3$. As was the case for SRD, the kinetic viscosity has no anti-symmetric component; the collisional contribution, however, does. Again, as discussed in Sect. 4.1.1 for SRD, one finds $v_1^{\text{col}} = v_2^{\text{col}} = v^{\text{col}}/2$. This relation is true for all $-a$ versions of MPC discussed in [32, 58, 59]. Simulation results were found to be in good agreement with theory.

For MPC-AT$+a$ it was found for sufficiently large M that [38, 59]

$$v^{\text{kin}} = \frac{k_B T \Delta t}{m} \left[\frac{M}{M - (d+2)/4} - \frac{1}{2} \right],$$

$$v^{\text{col}} = \frac{a^2}{24 \Delta t} \left(\frac{M - 7/5}{M} \right). \tag{58}$$

MPC-AT$-a$ and MPC-AT$+a$ both have the same kinetic contribution to the viscosity in two dimensions; however, imposing angular-momentum conservation makes the collisional contribution to the stress tensor symmetric, so that the asymmetric contribution, v_2, discussed in Sect. 4.1.1 vanishes. The resulting collisional contribution to the viscosity is then reduced by a factor close to 2.

6 Generalized MPC Algorithms for Dense Liquids and Binary Mixtures

The original SRD algorithm models a single-component fluid with an ideal-gas equation of state. The fluid is therefore very compressible, and the speed of sound, c_s, is low. In order to have negligible compressibility effects, as in real liquids, the Mach number has to be kept small, which means that there are limits on the flow velocity in the simulation. The SRD algorithm can be modified to model both excluded volume effects, allowing for a more realistic modeling of dense gases and liquids, as well as repulsive hard-core interactions between components in mixtures, which allow for a thermodynamically consistent modeling of phase separating mixtures.

6.1 Non-Ideal Model

As in SRD, the algorithm consists of individual streaming and collision steps. In order to define the collisions, a second grid with sides of length $2a$ is introduced, which (in $d = 2$) groups four adjacent cells into one "supercell." The cell structure is sketched in Fig. 3 (left panel). To initiate a collision, pairs of cells in every supercell are chosen at random. Three different choices are possible: (a) horizontal (with $\sigma_1 = \hat{x}$), (b) vertical ($\sigma_2 = \hat{y}$), and (c) diagonal collisions (with $\sigma_3 = (\hat{x} + \hat{y})/\sqrt{2}$ and $\sigma_4 = (\hat{x} - \hat{y})/\sqrt{2}$).

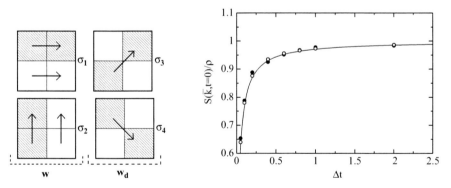

Fig. 3 *Left panel:* Schematic of collision rules. Momentum is exchanged in three ways: (a) horizontally along σ_1, (b) vertically along σ_2, and (c) diagonally along σ_3 and σ_4. w and w_d denote the probabilities of choosing collisions (a), (b), and (c), respectively. *Right panel:* Static structure factor $S(\bar{k}, t = 0)$ as a function of Δt for $M = 3$. The *open circles* show results obtained by taking the numerical derivative of the pressure. The *filled circles* are data obtained from direct measurements of the density fluctuations. The *solid line* is the theoretical prediction obtained using the first term in (61) and (63). $\bar{k}$ is the smallest wave vector, $\bar{k} = (2\pi/L)(1,0)$. Parameters: $L/a = 32$, $A = 1/60$, and $k_B T = 1.0$. From [33]

For a mean particle velocity $\mathbf{u}_n = (1/M_n) \sum_{i=1}^{M_n} \mathbf{v}_i$, of cell n, the projection of the difference of the mean velocities of the selected cell pairs on σ_j, $\Delta u = \sigma_j \cdot (\mathbf{u}_1 - \mathbf{u}_2)$, is then used to determine the probability of collision. If $\Delta u < 0$, no collision will be performed. For positive Δu, a collision will occur with an acceptance probability, p_A, which depends on Δu and the number of particles in the two cells, N_1 and N_2. The choice of p_A determines both the equation of state and the values of the transport coefficients. While there is considerable freedom in choosing p_A, the requirement of thermodynamic consistency imposes certain restrictions [33, 34, 55]. One possible choice is

$$p_A(M_1, M_2, \Delta u) = \Theta(\Delta u) \tanh(\Lambda) \quad \text{with} \quad \Lambda = A \Delta u N_1 N_2, \tag{59}$$

where Θ is the unit step function and A is a parameter which is used to tune the equation of state. The choice $\Lambda \sim N_1 N_2$ leads to a non-ideal contribution to the pressure which is quadratic in the particle density.

The collision rule chosen in [33] maximizes the momentum transfer parallel to the connecting vector σ_j and does not change the transverse momentum. It exchanges the parallel component of the mean velocities of the two cells, which is equivalent to a "reflection" of the relative velocities, $v_i^{\|}(t + \Delta t) - u^{\|} = -(v_i^{\|}(t) - u^{\|})$, where $u^{\|}$ is the parallel component of the mean velocity of the particles of *both* cells. This rule conserves momentum and energy in the cell pairs.

Because of $x - y$ symmetry, the probabilities for choosing cell pairs in the x- and y-directions (with unit vectors σ_1 and σ_2 in Fig. 3) are equal, and will be denoted by w. The probability for choosing diagonal pairs (σ_3 and σ_4 in Fig. 3) is given by $w_d = 1 - 2w$. w and w_d must be chosen so that the hydrodynamic equations are isotropic and do not depend on the orientation of the underlying grid. An equivalent criterion is to guarantee that the relaxation of the velocity distribution is isotropic. These conditions require $w = 1/4$ and $w_d = 1/2$. This particular choice also ensures that the kinetic part of the viscous stress tensor is isotropic [45].

6.1.1 Transport Coefficients

The transport coefficients can be determined using the same GK formalism as was used for the original SRD algorithm [21, 51]. Alternatively, the non-equilibrium approach described in Sect. 5 can be used. Assuming molecular chaos and ignoring fluctuations in the number of particles per cell, the kinetic contribution to the viscosity is found to be

$$\nu^{\text{kin}} = \frac{k_B T}{m} \Delta t \left(\frac{1}{p_{\text{col}}} - \frac{1}{2} \right) \quad \text{with} \quad p_{\text{col}} = A \sqrt{\frac{k_B T}{m\pi}} M^{3/2}, \tag{60}$$

which is in good agreement with simulation data. p_{col} is essentially the collision rate, and can be obtained by averaging the acceptance probability, (59). The collisional contribution to the viscosity is $\nu^{\text{col}} = p_{\text{col}}(a^2/3\Delta t)$ [60]. The self-diffusion constant,

D, is evaluated by summing over the velocity-autocorrelation function (see, e.g., [21]); which yields $D = \nu_{\mathrm{kin}}$.

6.1.2 Equation of State

The collision rules conserve the kinetic energy, so the internal energy should be the same as that of an ideal gas. Thermodynamic consistency therefore requires that the non-ideal contribution to the pressure is linear in T. This is possible if the coefficient A in (59) is sufficiently small.

The mechanical definition of pressure – the average longitudinal momentum transfer across a fixed interface per unit time and unit surface area – can be used to determine the equation of state. Only the momentum transfer due to collisions needs to be considered, since that coming from streaming constitutes the ideal part of the pressure. Performing this calculation for a fixed interface and averaging over the position of the interface, one finds the non-ideal part of the pressure,

$$P_{\mathrm{n}} = \left(\frac{1}{2\sqrt{2}} + \frac{1}{4} \right) \frac{A M^2}{2} \frac{k_{\mathrm{B}} T}{a \Delta t} + O(A^3 T^2). \tag{61}$$

P_{n} is quadratic in the particle density, $\rho = M/a^2$, as would be expected from a virial expansion. The prefactor A must be chosen small enough that higher-order terms in this expansion are negligible. Prefactors A leading to acceptance rates of about 15% are sufficiently small to guarantee that the pressure is linear in T.

The total pressure is the average of the diagonal part of the microscopic stress tensor,

$$P = P_{\mathrm{id}} + P_{\mathrm{n}} = \frac{1}{\Delta t L_x L_y} \left\langle \sum_j \left\{ \Delta t v_{jx}^2 - \Delta v_{jx} z_{jlx}^s / 2 \right\} \right\rangle. \tag{62}$$

The first term gives the ideal part of the pressure, P_{id}, as discussed in [21]. The average of the second term is the non-ideal part of the pressure, P_{n}. $\mathbf{z}_{jl}^s$ is a vector which indexes collision partners. The first subscript denotes the particle number and the second, l, is the index of the collision vectors $\boldsymbol{\sigma}_l$ in Fig. 3 (left panel). The components of $\mathbf{z}_{jl}^s$ are either 0, 1, or -1 [55]. Simulation results for P_{n} obtained using (62) are in good agreement with the analytical expression, (61). In addition, measurements of the static structure factor $S(k \to 0, t = 0)$ agree with the thermodynamic prediction

$$S(k \to 0, t = 0) = \rho k_{\mathrm{B}} T \partial \rho / \partial P|_T \tag{63}$$

when result (61) is used [see Fig. 3 (right panel)]. The adiabatic speed of sound obtained from simulations of the dynamic structure factor is also in good agreement with the predictions following from (61). These results provide strong evidence for the thermodynamic consistency of the model. Consistency checks are particularly important because the non-ideal algorithm does not conserve phase-space volume. This is because the collision probability depends on the difference of collision-cell velocities, so that two different states can be mapped onto the same state by a collision. While the dynamics presumably still obeys detailed – or at least semi-detailed – balance, this is very hard to prove, since it would require knowledge not only of the transition probabilities, but also of the probabilities of the individual equilibrium states. Nonetheless, no inconsistencies due to the absence of time-reversal invariance or a possible violation of detailed balance have been observed.

The structure of $S(k)$ for this model is also very similar to that of a simple dense fluid. In particular, for fixed M, both the depth of the minimum at small k and the height of the first peak increase with decreasing Δt, until there is an order–disorder transition. The fourfold symmetry of the resulting ordered state – in which clusters of particles are concentrated at sites with the periodicity close, but not necessarily equal, to that of the underlying grid – is clearly dictated by the structure of the collision cells. Nevertheless, these ordered structures are similar to the low-temperature phase of particles with a strong repulsion at intermediate distances, but a soft repulsion at short distances. The scaling behavior of both the self-diffusion constant and the pressure persists until the order/disorder transition.

6.2 Phase-Separating Multi-Component Mixtures

In a binary mixture of A and B particles, phase separation can occur when there is an effective repulsion between A–B pairs. In the current model, this is achieved by introducing velocity-dependent multi-particle collisions between A and B particles. There are N_A and N_B particles of type A and B, respectively. In two dimensions, the system is coarse-grained into $(L/a)^2$ cells of a square lattice of linear dimension L and lattice constant a. The generalization to three dimensions is straightforward.

Collisions are defined in the same way as in the non-ideal model discussed in the previous section. Now, however, two types of collisions are possible for each pair of cells: particles of type A in the first cell can undergo a collision with particles of type B in the second cell; vice versa, particles of type B in the first cell can undergo a collision with particles of type A in the second cell. There are no A–A or B–B collisions, so that there is an effective repulsion between A–B pairs. The rules and probabilities for these collisions are chosen in the same way as in the non-ideal single-component fluid described in [33, 55]. For example, consider the collision of A particles in the first cell with the B particles in the second. The mean particle velocity of A particles in the first cell is $\mathbf{u}_A = (1/N_{c,A}) \sum_{i=1}^{N_{c,A}} \mathbf{v}_i$, where the sum runs over all A particles, $N_{c,A}$, in the first cell. Similarly, $\mathbf{u}_B = (1/N_{c,B}) \sum_{i=1}^{N_{c,B}} \mathbf{v}_i$ is the mean velocity of B particles in the second cell. The projection of the difference of the mean velocities of the selected cell-pairs on σ_j, $\Delta u_{AB} = \sigma_j \cdot (\mathbf{u}_A - \mathbf{u}_B)$, is then used to determine the probability of collision. If $\Delta u_{AB} < 0$, no collision will be performed. For positive Δu_{AB}, a collision will occur with an acceptance probability

$$p_A(N_{c,A}, N_{c,B}, \Delta u_{AB}) = A \, \Delta u_{AB} \, \Theta(\Delta u_{AB}) N_{c,A} N_{c,B}, \qquad (64)$$

where Θ is the unit step function and A is a parameter which allows us to tune the equation of state; in order to ensure thermodynamic consistency, it must be sufficiently small that $p_A < 1$ for essentially all collisions. When a collision occurs, the parallel component of the mean velocities of colliding particles in the two cells, $v_i^\parallel(t + \Delta t) - u_{AB}^\parallel = -(v_i^\parallel(t) - u_{AB}^\parallel)$, is exchanged, where $u_{AB}^\parallel = (N_{c,A} u_A^\parallel + N_{c,B} u_B^\parallel)/(N_{c,A} + N_{c,B})$ is the parallel component of the mean velocity of the colliding particles. The perpendicular component remains unchanged. It is easy to verify that these rules conserve momentum and energy in the cell pairs. The collision of B particles in the first cell with A particles in the second is handled in a similar fashion.

Because there are no A–A and B–B collisions, additional SRD collisions at the cell level are incorporated in order to mix particle momenta. The order of A–B and SRD collision is random, i.e., the SRD collision is performed first with a probability $1/2$. If necessary, the viscosity can be tuned by not performing SRD collisions every time step. The results presented here were obtained using a SRD collision angle of $\alpha = 90°$.

The transport coefficients can be calculated in the same way as for the one-component non-ideal system. The resulting kinetic contribution to the viscosity is

$$v^{kin} = \frac{\Delta t k_B T}{2} \left\{ \frac{1}{A} \sqrt{\frac{2\pi}{k_B T}} [M_A M_B (M_A + M_B)]^{-1/2} - 1 \right\}, \tag{65}$$

where $M_A = \langle N_{c,A} \rangle$, $M_B = \langle N_{c,B} \rangle$. In deep quenches, the concentration of the minority component is very small, and the non-ideal contribution to the viscosity approaches zero. In this case, the SRD collisions provide the dominant contribution to the viscosity.

6.2.1 Free Energy

An analytic expression for the equation of state of this model can be derived by calculating the momentum transfer across a fixed surface, in much the same way as was done for the non-ideal model in [33]. Since there are only non-ideal collisions between A–B particles, the resulting contribution to the pressure is

$$P_n = \left(w + \frac{w_d}{\sqrt{2}} \right) A M_A M_B \frac{k_B T}{a \Delta t} = \Gamma \rho_A \rho_B, \tag{66}$$

where ρ_A and ρ_B are the densities of A and B and $\Gamma \equiv (w + w_d/\sqrt{2}) a^3 A / \Delta t$. In simulations, the total pressure can be measured by taking the ensemble average of the diagonal components of the microscopic stress tensor. In this way, the pressure can be measured locally, at the cell level. In particular, the pressure in a region consisting of N_{cell} cells is

$$P_n = \frac{1}{\Delta t a^2 N_{cell}} \left\langle \sum_{c=1}^{N_c} \sum_{j \in c} \left[\Delta t v_{jx}^2 - \Delta v_{jx} z_{jlx}^s / 2 \right] \right\rangle, \tag{67}$$

where the second sum runs over the particles in cell c. The first term in (67) is the ideal-gas contribution to the pressure; the second comes from the momentum transfer between cells involved in the collision indexed by z_{jl}^s [45].

Expression (66) can be used to determine the entropy density, s. The ideal-gas contribution to s has the form [61]

$$s_{ideal} = \rho \, \varphi(T) - k_B \left[\rho_A \ln \rho_A + \rho_B \ln \rho_B \right], \tag{68}$$

where $\rho = \rho_A + \rho_B$. Since $\varphi(T)$ is independent of ρ_A and ρ_B, this term does not play a role in the current discussion. The non-ideal contribution to the entropy density, s_n, can be obtained from (66) using the thermodynamic relation

$$P_n/T = -s_n + \rho_A \partial s_n / \partial \rho_A + \rho_B \partial s_n / \partial \rho_B. \tag{69}$$

The result is $s_n = \Gamma \rho_A \rho_B$, so that the total configurational contribution to the entropy density is

$$s = -k_B \left\{ \rho_A \ln \rho_A + \rho_B \ln \rho_B + \Gamma \rho_A \rho_B \right\}. \tag{70}$$

Since there is no configurational contribution to the internal energy in this model, the mean-field phase diagram can be determined by maximizing the entropy at fixed density ρ. The resulting demixing phase diagram as a function of $\rho_{AB} = (\rho_A - \rho_B)/\rho$ is given by the solid line in Fig. 4 (left panel). The critical point is located at $\rho_{AB} = 0$, $\rho \Gamma^* = 2$. For $\rho \Gamma < 2$, the order parameter $\rho_{AB} = 0$; for $\rho \Gamma > 2$, there is phase separation into coexisting A- and B-rich phases. As can be seen, the agreement between the mean-field predictions and simulation results is very good except close to the critical point, where the histogram method of determining the coexisting densities is unreliable and critical fluctuations influence the shape of the coexistence curve.

6.2.2 Surface Tension

A typical configuration for $\rho_{AB} = 0$, $\rho \Gamma = 3.62$ is shown in the inset to Fig. 4 (left panel), and a snapshot of a fluctuating droplet at $\rho_{AB} = -0.6$, $\rho \Gamma = 3.62$ is shown in the inset to Fig. 4 (right panel). The amplitude of the capillary wave fluctuations of a droplet is determined by the surface tension, σ. Using the parameterization $r(\phi) = r_0 \left[1 + \sum_{k=-\infty}^{\infty} u_k \exp(ik\phi) \right]$ and choosing u_0 to fix the area of the droplet, it can be shown that [54]

$$\langle |u_k|^2 \rangle = \frac{k_B T}{2\pi r_0 \sigma} \left(\frac{1}{k^2 - 1} \right). \tag{71}$$

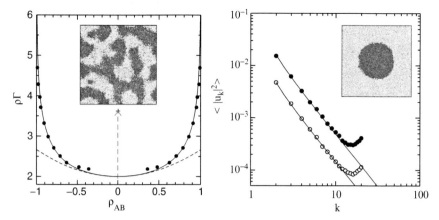

Fig. 4 *Left panel:* Binary phase diagram. There is phase separation for $\rho\Gamma > 2$. Simulation results for ρ_{AB} obtained from concentration histograms are shown as *filled circles.* The *dashed line* is a plot of the leading singular behavior, $\rho_{AB} = \sqrt{3(\rho\Gamma - 2)/2}$, of the order parameter at the critical point. The *inset* shows a configuration 50,000 time steps after a quench along $\rho_{AB} = 0$ to $\rho\Gamma = 3.62$ (*arrow*). The *dark (blue)* and *light (white) spheres* are A and B particles, respectively. Parameters: $L/a = 64$, $M_A = M_B = 5$, $k_B T = 0.0004$, $\Delta t = 1$, and $a = 1$. From [45]. *Right panel:* Dimensionless radial fluctuations, $\langle|u_k^2|\rangle$, as a function of the mode number k for $A = 0.45$ (*filled circles*) and $A = 0.60$ (*open circles*) with $k_B T = 0.0004$. The average droplet radii are $r_0 = 11.95a$ and $r_0 = 15.21a$, respectively. The *solid lines* are fits to (71). The *inset* shows a typical droplet configuration for $\rho_{AB} = -0.6$, $\rho\Gamma = 3.62$ ($A = 0.60$ and $k_B T = 0.0004$). Parameters: $L/a = 64$, $M_A = 2$, $M_B = 8$, $\Delta t = 1$, and $a = 1$. From [45]

Figure 4 (right panel) contains a plot of $\langle|u_k|^2\rangle$ as a function of mode number k for $\rho\Gamma = 3.62$ and $\rho\Gamma = 2.72$. Fits to the data yield $\sigma \simeq 2.9 k_B T$ for $\rho\Gamma = 3.62$ and $\sigma \simeq 1.1 k_B T$ for $\rho\Gamma = 2.72$. Mechanical equilibrium requires that the pressure difference across the interface of a droplet satisfies the Laplace equation

$$\Delta p = p_{\text{in}} - p_{\text{out}} = (d - 1)\sigma/r_0 \tag{72}$$

in d spatial dimensions. Measurements of Δp [using (67)] as a function of the droplet radius for $A = 0.60$ at $k_B T = 0.0005$ yield results in excellent agreement with the Laplace equation for the correct value of the surface tension [45].

The model therefore displays the correct thermodynamic behavior and interfacial fluctuations. It can also be extended to model amphiphilic mixtures by introducing dimers consisting of tethered A and B particles. If the A and B components of the dimers participate in the same collisions as the solvent, they behave like amphiphilic molecules in binary oil–water mixtures. The resulting model displays a rich phase behavior as a function of $\rho\Gamma$ and the number of dimers, N_d. Both the formation of droplets and micelles, as shown in Fig. 5 (left panel), and a bicontinuous phase, as illustrated in Fig. 5 (right panel), have been observed [45]. The coarse-grained nature

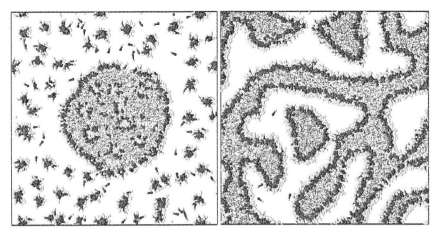

Fig. 5 *Left panel:* Droplet configuration in a mixture with $N_A = 8{,}192$, $N_B = 32{,}768$, and $N_d = 1{,}500$ dimers after 10^5 time steps. The initial configuration is a droplet with a homogeneous distribution of dimers. The *dark (blue)* and *light (white) colored spheres* indicate A and B particles, respectively. For clarity, A particles in the bulk are smaller and B particles in the bulk are not shown. Parameters: $L/a = 64$, $M_A = 2$, $M_B = 8$, $A = 1.8$, $k_B T = 0.0001$, $\Delta t = 1$, and $a = 1$. *Right panel:* Typical configuration showing the bicontinuous phase for $N_A = N_B = 20{,}480$ and $N_d = 3{,}000$. Parameters: $L/a = 64$, $M_A = 5$, $M_B = 5$, $A = 1.8$, $k_B T = 0.0001$, $\Delta t = 1$, and $a = 1$. From [45]

of the algorithm therefore enables the study of large time scales with a feasible computational effort.

6.2.3 Color Models for Immiscible Fluids

There have been other generalizations of SRD to model binary mixtures by Hashimoto et al. [62] and Inoue et al. [63], in which a color charge, $c_i = \pm 1$ is assigned to two different species of particles. The rotation angle α in the SRD rotation step is then chosen such that the color-weighted momentum in a cell, $\mathbf{m} = \sum_{i=1}^{N_c} c_i(\mathbf{v}_i - \mathbf{u})$, is rotated to point in the direction of the gradient of the color field $\bar{c} = \sum_{i=1}^{N_c} c_i$. This rule also leads to phase separation. Several tests of the model have been performed; Laplace's equation was verified numerically, and simulation studies of spinodal decomposition and the deformation of a falling droplet were performed [62]. Later applications include a study of the transport of slightly deformed immiscible droplets in a bifurcating channel [64]. Subsequently, the model was generalized through the addition of dumbbell-shaped surfactants to model micellization [65] and the behavior of ternary amphiphilic mixtures in both two and three dimensions [66, 67]. Note that since the color current after the collision is always parallel to the color gradient, thermal fluctuations of the order parameter are neglected in this approach.

7 Boundary Conditions and Embedded Objects

7.1 Collisional Coupling to Embedded Particles

A very simple procedure for coupling embedded objects such as colloids or polymers to a MPC solvent has been proposed in [68]. In this approach, every colloid particle or monomer in the polymer chain is taken to be a point-particle which participates in the SRD collision. If monomer i has mass m_m and velocity $\mathbf{w}_i$, the center of mass velocity of the particles in the collision cell is

$$\mathbf{u} = \frac{m \sum_{i=1}^{N_c} \mathbf{v}_i + m_m \sum_{i=1}^{N_m} \mathbf{w}_i}{N_c m + N_m m_m}, \tag{73}$$

where N_m is the number of monomers in the collision cell. A stochastic collision of the relative velocities of both the solvent particles and embedded monomers is then performed in the collision step. This results in an exchange of momentum between the solvent and embedded monomers. The same procedure can of course be employed for other MPC algorithms, such as MPC-AT. The new monomer momenta are then used as initial conditions for a molecular-dynamics update of the polymer degrees of freedom during the subsequent streaming time step, Δt. Alternatively, the momentum exchange, Δp, can be included as an additional force $\Delta p / \Delta t$ in the molecular-dynamics integration. If there are no other interactions between monomers – as might be the case for embedded colloids – these degrees of freedom stream freely during this time interval.

When using this approach, the average mass of solvent particles per cell, $m N_c$, should be of the order of the monomer or colloid mass m_m (assuming one embedded particle per cell) [15]. This corresponds to a neutrally buoyant object which responds quickly to the fluid flow but is not kicked around too violently. It is also important to note that the average number of monomers per cell, $\langle N_m \rangle$, should be smaller than unity in order to properly resolve hydrodynamic interactions between the monomers. On the other hand, the average bond length in a semi-flexible polymer or rod-like colloid should also not be much larger than the cell size a, in order to capture the anisotropic friction of rod-like molecules due to hydrodynamic interactions [69] (which leads to a twice as large perpendicular than parallel friction coefficient for long stiff rods [6]), and to avoid an unnecessarily large ratio of the number of solvent to solute particles. For a polymer, the average bond length should therefore be of the order of a.

In order to use SRD to model suspended colloids with a radius of order 1 μm in water, this approach would require approximately 60 solvent particles per cell in order to match the Peclet number [16]. This is much larger than the optimum number (see discussion in Sect. 2.2.1), and the relaxation to the Boltzmann distribution is very slow. Because of its simplicity and efficiency, this monomer–solvent coupling has been used in many polymer [14, 71–74] and colloid simulations [15, 16, 75, 76].

7.2 Thermal Boundaries

In order to accurately resolve the local flow field around a colloid, methods have been proposed which exclude fluid-particles from the interior of the colloid and mimic slip [19, 77] or no-slip [78] boundary conditions. The latter procedure is similar to what is known in molecular dynamics as a "thermal wall" boundary condition: fluid particles which hit the colloid particle are given a new, random velocity drawn from the following probability distributions for the normal velocity component, v_N, and the tangential component, v_T,

$$p_N(v_N) = (mv_N/k_BT) \exp\left(-mv_N^2/2k_BT\right), \quad \text{with} \quad v_N > 0,$$
$$p_T(v_T) = \sqrt{m/2\pi k_BT} \exp\left(-mv_T^2/2k_BT\right). \tag{74}$$

These probability distributions are constructed so that the probability distribution for particles near the wall remains Maxwellian. The probability distribution, p_T, for the tangential components of the velocity is Maxwellian, and both positive and negative values are permitted. The normal component must be positive, since after scattering at the surface, the particle must move away from the wall. The form of p_N is a reflection of the fact that more particles with large $|v_N|$ hit the wall per unit time than with small $|v_N|$ [78].

This procedure models a no-slip boundary condition at the surface of the colloid, and also thermostats the fluid at the boundaries. For many non-equilibrium flow conditions, this may not be sufficient, and it may also be necessary to thermostat the bulk fluid also (compare Sect. 2.3). It should also be noted that (74) will be a good approximation only if the radius of the embedded objects is much larger than the mean free path λ. For smaller particles, corrections are needed.

If a particle hits the surface at time t_0 in the interval between $n\Delta t$ and $(n+1)\Delta t$, the correct way to proceed would be to give the particle its new velocity and then have it stream the remaining time $(n+1)\Delta t - t_0$. However, such detailed resolution is not necessary. It has been found [16] that good results are also obtained using the following simple stochastic procedure. If a particle is found to have penetrated the colloid during the streaming step, one simply moves it to the boundary and then stream a distance $\mathbf{v}_{\text{new}} \Delta t \epsilon$, where ϵ is a uniformly distributed random number in the interval $[0,1]$.

Another subtlety is worth mentioning. If two colloid particles are very close, it can happen that a solvent particle could hit the second colloid after scattering off the first, all in the interval Δt. Naively, one might be tempted to simply forbid this from happening or ignore it. However, this would lead to a strong depletion-like attractive force between the colloids [16]. This effect can be greatly reduced by allowing multiple collisions in which one solvent particle is repeatedly scattered off the two colloids. In every collision, momentum is transferred to one of the colloids, which pushes the colloids further apart. In practice, even allowing for up to ten multiple collisions cannot completely cancel the depletion interaction – one needs an additional repulsive force to eliminate this unphysical attraction. The same effect can occur when a colloid particle is near a wall.

Careful tests of this thermal coupling have been performed by Padding et al. [17, 79], who were able to reproduce the correct rotational diffusion of a colloid. It should be noted that because the coupling between the solvent particles and the surface occurs only through the movement of the fluid particles, the coupling is quite weak for small mean free paths.

7.3   Coupling Using Additional Forces

Another procedure for coupling an embedded object to the solvent has been pursued by Kapral et al. [19, 30, 80]. They introduce a central repulsive force between the solvent particles and the colloid. This force has to be quite strong in order to prohibit a large number of solvent particles from penetrating the colloid. When implementing this procedure, a small time step δt is therefore required in order to resolve these forces correctly, and a large number of molecular dynamics time steps are needed during the SRD streaming step. In its original form, central forces were used, so that only slip boundary conditions could be modeled. In principle, non-central forces could be used to impose no-slip conditions.

This approach is quite natural and very easy to implement; it does, however, require the use of small time steps and therefore may not be the optimal procedure for many applications.

7.4   "Ghost" or "Wall" Particles

One of the first approaches employed to impose a non-slip boundary condition at an external wall or at a moving object in a MPC solvent was to use "ghost" or "wall" particles [36, 81]. In other mesoscale methods such as LB, no-slip conditions are modeled using the bounce-back rule: the velocity of the particle is inverted from $\mathbf{v}$ to $-\mathbf{v}$ when it intersects a wall. For planar walls which coincide with the boundaries of the collision cells, the same procedure can be used in MPC. However, the walls will generally not coincide with, or even be parallel to, the cell walls. Furthermore, for small mean free paths, where a shift of the cell lattice is required to guarantee Galilean invariance, partially occupied boundary cells are unavoidable, even in the simplest flow geometries.

The simple bounce-back rule fails to guarantee no-slip boundary conditions in the case of partially filled cells. The following generalization of the bounce-back rule has therefore been suggested. For all cells that are cut by walls, fill the "wall" part of the cell with a sufficient number of virtual particles in order to make the total number of particles equal to M, the average number of particles per cell. The velocities of the wall particles are drawn from a Maxwell–Boltzmann distribution with zero mean

Fig. 6 Velocity field of a fluid near a square cylinder in a Poiseuille flow at Reynolds number $Re = v_{\max}L/v = 30$. The channel width is eight times larger than the cylinder size L. A pair of stationary vortices is seen behind the obstacle, as expected for $Re \leq 60$. From [81]

velocity and the same temperature as the fluid. The collision step is then carried out using the mean velocity of all particles in the cell. Note that since Gaussian random numbers are used, and the sum of Gaussian random numbers is also Gaussian-distributed, the velocities of the individual wall particles need not be determined explicitly. Instead, the average velocity $\mathbf{u}$ can be written as $\mathbf{u} = \left(\sum_{i=1}^{n} \mathbf{v}_i + \mathbf{a}\right)/M$, where $\mathbf{a}$ is a vector whose components are Gaussian random numbers with zero mean and variance $(M - n)k_{\mathrm{B}}T$. Results for Poiseuille flow obtained using this procedure, both with and without cell shifting, were found to be in excellent agreement with the correct parabolic profile [36]. Similarly, numerical results for the recirculation length, the drag coefficient, and the Strouhal number for flows around a circular and square cylinder in two dimensions were shown to be in good agreement with experimental results and computational fluid dynamics data for a range of Reynolds numbers between $Re = 10$ and $Re = 130$ (see Fig. 6) [36, 81].

8 Importance of Angular-Momentum Conservation: Couette Flow

As an example of a situation in which it is important to use an algorithm which conserves angular momentum, consider a drop of a highly viscous fluid inside a lower-viscosity fluid in circular Couette flow. In order to avoid the complications of phase-separating two-component fluids, the high viscosity fluid is confined to a radius $r < R_1$ by an impenetrable boundary with reflecting boundary conditions (i.e., the momentum parallel to the boundary is conserved in collisions). No-slip boundary conditions between the inner and outer fluids are guaranteed because collision cells reach across the boundary. When a torque is applied to the outer circular wall (with no-slip, bounce-back boundary conditions) of radius $R_2 > R_1$, a solid-body rotation of *both fluids* is expected. The results of simulations with both MPC-AT$-a$

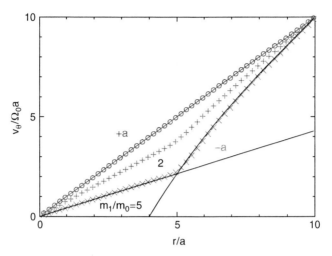

Fig. 7 Azimuthal velocity of binary fluids in a rotating cylinder with $\Omega_0 = 0.01(k_B T/m_0 a^2)^{1/2}$. The viscous fluids with particle mass m_1 and m_0 are located at $r < R_1$ and $R_1 < r < R_2$, respectively, with $R_1 = 5a$ and $R_2 = 10a$. Symbols represent the simulation results of MPC-AT$-a$ with $m_1/m_0 = 2$ (*pluses*) or $m_1/m_0 = 5$ (*crosses*), and MPC-AT$+a$ for $m_1/m_0 = 5$ (*open circles*). *Solid lines* represent the analytical results for MPC-AT$-a$ at $m_1/m_0 = 5$. Error bars are smaller than the size of the symbols. From [38]

and MPC-AT$+a$ are shown in Fig. 7. While MPC-AT$+a$ reproduces the expected behavior, MPC-AT$-a$ produces different angular velocities in the two fluids, with a low (high) angular velocity in the fluid of high (low) viscosity [38].

The origin of this behavior is that the viscous stress tensor in general has symmetric and antisymmetric contributions (see Sect. 4.1.1),

$$\sigma_{\alpha\beta} = \lambda(\partial_\gamma v_\gamma)\delta_{\alpha\beta} + \bar{\eta}\left(\partial_\beta v_\alpha + \partial_\alpha v_\beta\right) + \check{\eta}\left(\partial_\beta v_\alpha - \partial_\alpha v_\beta\right), \qquad (75)$$

where λ is the second viscosity coefficient and $\bar{\eta} \equiv \rho v_1$ and $\check{\eta} \equiv \rho v_2$ are the symmetric and anti-symmetric components of the viscosity, respectively. The last term in (75) is linear in the vorticity $\nabla \times \mathbf{v}$, and does not conserve angular momentum. This term therefore vanishes (i.e., $\check{\eta} = 0$) when angular momentum is conserved.

The anti-symmetric part of the stress tensor implies an additional torque, which becomes relevant when the boundary condition is given by forces. In cylindrical coordinates (r, θ, z), the azimuthal stress is given by [38]

$$\sigma_{r\theta} = (\bar{\eta} + \check{\eta})\frac{r\partial(v_\theta/r)}{\partial r} + 2\check{\eta}\frac{v_\theta}{r}. \qquad (76)$$

The first term is the stress of the angular-momentum-conserving fluid, which depends on the derivative of the angular velocity $\Omega = v_\theta/r$. The second term is an additional stress caused by the lack of angular momentum conservation; it is proportional to Ω.

In the case of the phase-separated fluids in circular Couette flow, this implies that if both fluids rotate at the same angular velocity, the inner and outer stresses do not coincide. Thus, the angular velocity of the inner fluid Ω_1 is smaller than the outer one, with $v_\theta(r) = \Omega_1 r$ for $r < R_1$ and

$$v_\theta(r) = Ar + B/r, \quad \text{with} \quad A = \frac{\Omega_2 R_2^2 - \Omega_1 R_1^2}{R_2^2 - R_1^2}, \quad B = \frac{(\Omega_1 - \Omega_2)R_1^2 R_2^2}{R_2^2 - R_1^2} \quad (77)$$

for $R_1 < r < R_2$. Ω_1 is then obtained from the stress balance at $r = R_1$, i.e., $2\check{\eta}_1 \Omega_1 = (8/3)\eta_2(\Omega_0 - \Omega_1) + 2\check{\eta}_2 \Omega_1$. This calculation reproduces the numerical results very well, see Fig. 7. Thus, it is essential to employ an $+a$ version of MPC in simulations of multi-phase flows of binary fluids with different viscosities.

There are other situations in which the lack of angular momentum conservation can cause significant deviations. In [38], a star polymer with small monomer spacing was placed in the middle of a rotating Couette cell. As in the previous case, it was observed that the polymer fluid rotated with a smaller angular velocity than the outer fluid. When the angular momentum conservation was switched on, everything rotated at the same angular velocity, as expected.

9 MPC without Hydrodynamics

The importance of hydrodynamic interactions (HI) in complex fluids is generally accepted. A standard procedure for determining the influence of HI is to investigate the same system with and without HI. In order to compare results, however, the two simulations must differ as little as possible – apart from the inclusion of HI. A well-known example of this approach is Stokesian dynamics simulations (SD), where the original BD method can be extended by including hydrodynamic interactions in the mobility matrix by employing the Oseen tensor [6, 12].

A method for switching off HI in MPC has been proposed in [24, 26]. The basic idea is to randomly interchange velocities of all solvent particles after each collision step, so that momentum (and energy) are *not* conserved *locally*. Hydrodynamic correlations are therefore destroyed, while leaving friction coefficients and fluid self-diffusion coefficients largely unaffected. Since this approach requires the same numerical effort as the original MPC algorithm, a more efficient method has been suggested recently in [25]. If the velocities of the solvent particles are not correlated, it is no longer necessary to follow their trajectories. In a random solvent, the solvent-solute interaction in the collision step can thus be replaced by the interaction with a heat bath. This strategy is related to that proposed in [36] to model no-slip boundary conditions of solvent particles at a planar wall, compare Sect. 7.4. Since the positions of the solvent particles within a cell are not required in the collision step, no explicit particles have to be considered. Instead, each monomer is coupled with an effective solvent momentum $\mathbf{P}$ which is directly chosen from a Maxwell–Boltzmann distribution of variance mMk_BT and a mean given by the average momentum of the fluid field – which is zero at rest, or $(mM\dot{\gamma}r_y^i, 0, 0)$ in the case of an imposed shear flow. The total center-of-mass velocity, which is used in the collision step, is then given by [25]

$$\mathbf{v}_{\mathrm{cm},i} = \frac{m_{\mathrm{m}}\mathbf{v}_i + \mathbf{P}}{mM + m_{\mathrm{m}}}, \tag{78}$$

where m_{m} is the mass of the solute particle. The solute trajectory is then determined using MD, and the interaction with the solvent is performed every collision time Δt.

The random MPC solvent therefore has similar properties to the MPC solvent, except that there are no HI. The relevant parameters in both methods are the average number of particles per cell, M, the rotation angle α, and the collision time Δt which can be chosen to be the same. For small values of the density ($M < 5$), fluctuation effects have been noticed [26] and could also be included in the random MPC solvent by a Poisson-distributed density. The velocity autocorrelation functions [15] of a random MPC solvent show a simple exponential decay, which implies some differences in the solvent diffusion coefficients. Other transport coefficients such as the viscosity depend on HI only weakly [57] and consequently are expected to be essentially identical in both solvents.

10 Applications to Colloid and Polymer Dynamics

The relevance of hydrodynamic interactions for the dynamics of complex fluids – such as dilute or semidilute polymer solutions, colloid suspensions, and microemulsions – is well known [6, 12]. From the simulation point of view, however, these systems are difficult to study because of the large gap in length- and time-scales between solute and solvent dynamics. One possibility for investigating complex fluids is the straightforward application of molecular dynamics simulations (MD), in which the fluid is course-grained and represented by Lennard-Jones particles. Such simulations provide valuable insight into polymer dynamics [83–87]. Similarly, mesoscale algorithms such as LB and DPD have been widely used for modeling of colloidal and polymeric systems [88–92].

Solute molecules, e.g., polymers, are typically composed of a large number of individual particles, whose interactions are described by a force-field. As discussed in Sect. 7, the particle-based character of the MPC solvent allows for an easy and controlled coupling between the solvent and solute particles. Hybrid simulations combining MPC and molecular dynamics simulations are therefore easy to implement. Results of such hybrid simulations are discussed in the following.

10.1 Colloids

Many applications in chemical engineering, geology, and biology involve systems of particles immersed in a liquid or gas flow. Examples include sedimentation processes, liquid-solid fluidized beds, and flocculation in suspensions.

Long-range solvent-mediated hydrodynamic interactions have a profound effect on the non-equilibrium properties of colloidal suspensions, and the many-body hydrodynamic backflow effect makes it difficult to answer even relatively simple questions such as what happens when a collection of particles sediments through a viscous fluid. Batchelor [93] calculated the lowest-order volume fraction correction to the average sedimentation velocity, $v_s = v_s^0(1 - 6.55\,\phi)$, of hard spheres of hydrodynamic radius R_H where v_s^0 is the sedimentation velocity of a single sphere. Because of the complicated interplay between short-range contact forces and long-range HI, it is difficult to extend this result to the high volume fraction suspensions of interest for ceramics and soil mechanics. An additional complication is that the Brownian motion of solute particles in water cannot be neglected if they are smaller than 1 µm in diameter.

The dimensionless Peclet number $Pe = v_s^0 R_H/D$, where D is the self-diffusion coefficient of the suspended particles, measures the relative strength of HI and thermal motion. Most studies of sedimentation have focused on the limit of infinite Peclet number, where Brownian forces are negligible. For example, Ladd [94] employed a LB method, and Höfler and Schwarzer [95] used a marker-and-cell Navier–Stokes solver to simulate such non-Brownian suspensions. The main difficulty with such algorithms is the solid-fluid coupling which can be very tricky: in LB simulations, special "boundary nodes" were inserted on the colloid surface, while in [95], the coupling was mediated by inertia-less markers which are connected to the colloid by stiff springs and swim in the fluid, effectively dragging the colloid, but also exerting a force on the fluid. Several methods for coupling embedded particles to an MPC solvent were discussed in Sect. 7.

Using the force-based solvent-colloid coupling described in Sect. 7.3, Padding and Louis [96] investigated the importance of HI during sedimentation at small Peclet numbers. Surprisingly, they found that the sedimentation velocity does not change if the Peclet number is varied between 0.1 and 15 for a range of volume fractions. For small volume fractions, the numerical results agree with the Batchelor law; for intermediate ϕ they are consistent with the semi-empirical Richardson–Zaki law, $v_s = v_s^0(1 - \phi)^n$, $n = 6.55$. Even better agreement was found with theoretical predictions by Hayakawa and Ichiki [97, 98], who took higher-order HI into account. Purely hydrodynamic arguments are therefore still valid in an average sense at low Pe, i.e., for strong Brownian motion and relatively weak HI. This also means that pure Brownian simulations without HI, which lead to $v_s = v_s^0(1 - \phi)$, strongly underestimate the effect of backflow.

On the other hand, it is known that the velocity autocorrelation function of a colloidal particle embedded in a fluctuating liquid at equilibrium exhibits a hydrodynamic long-time tail, $\langle v(t)v(0)\rangle \sim t^{-d/2}$, where d is the spatial dimension [99]. These tails have been measured earlier for point-like SRD particles in two [21, 27] and three [15] spatial dimensions, and found to be in quantitative agreement with analytic predictions, with no adjustable parameters. It is therefore not surprising that good agreement was also obtained for embedded colloids [96]. MPC therefore correctly describes two of the most important effects in colloidal suspensions, thermal fluctuations and hydrodynamic interactions.

In a series of papers, Hecht et al. [16, 76, 100] used hybrid SRD–MD simulations to investigate a technologically important colloidal system – Al_2O_3-particles of diameter 0.5 μm (which is often used in ceramics) suspended in water – with additional colloid–colloid interactions. These colloids usually carry a charge which, by forming an electric double layer with ions in water, results in a screened electrostatic repulsion. The interaction can be approximated by the Derjaguin–Landau–Verwey–Overbeek (DLVO) theory [101, 102]. The resulting potential contains a repulsive Debye–Hückel contribution, $V_{EL} \sim \exp(-\kappa[r-d])/r$, where d is the particle diameter, κ is the inverse screening length, and r is the distance of the particle centers. The second part of the DLVO-potential is a short-range van der Waals attraction,

$$V_{vdW} = -\frac{A_H}{12}\left[\frac{d^2}{r^2-d^2}+\frac{d^2}{r^2}+2\ln\left(\frac{r^2-d^2}{r^2}\right)\right],\qquad(79)$$

which turns out to be important at the high volume fractions ($\phi > 20\%$) and high salt concentrations of interest. A_H is the Hamaker constant which involves the polarizability of the particles and the solvent. DLVO theory makes the assumption of linear polarizability and is valid only at larger distances. It therefore does not include the so-called primary potential minimum at particle contact, which is observed experimentally and is about $30 k_B T$ deep. Because of this potential minimum, colloids which come in contact rarely become free again. In order to ensure numerical stability for reasonable values of the time step, this minimum was modeled by an additional parabolic potential with depth of order $6 k_B T$. The particle Reynolds number of the real system is very small, of order 10^{-6}–10^{-7}. Since it would be too time-consuming to model this Reynolds number, the simulations were performed at $Re \approx 0.02$, which still ensures that the contribution of momentum convection is negligible compared to that of momentum diffusion. However, due to the remaining inertial effects and the non-zero time step, it was still possible that particles partially overlapped in the simulation. This overlap was penalized by an additional potential, frequently used in simulations of granular matter, given by a Hertz-law,

$$V_{Hertz} \sim (d-r)^{5/2} \quad \text{if } r < d.\qquad(80)$$

SRD correctly describes long-range HI, but it can only resolve hydrodynamic interactions on scales larger than both the mean free path λ and the cell size a. In a typical simulation with about 1,000 colloid particles, a relatively small colloid diameter of about four lattice units was chosen for computational efficiency. This means that HI are not fully resolved at interparticle distances comparable to the colloid diameter, and lubrication forces have to be inserted by hand. Only the most divergent mode, the so-called squeezing mode, was used, $F_{lub} \sim v_{rel}/r_{rel}$, where r_{rel} and v_{rel} are the relative distance and velocity of two colloids, respectively. This system of interacting Al_2O_3-particles was simulated in order to study the dependence of the suspension's viscosity and structure on shear rate, pH, ionic strength and volume fraction. The resulting stability diagram of the suspension as a function of ionic strength and pH value is shown in Fig. 8 (plotted at zero shear) [76]. The pH

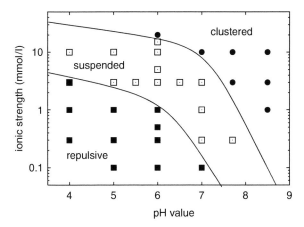

Fig. 8 Phase diagram of a colloidal suspension (plotted at zero shear and volume fraction $\phi = 35\%$) in the ionic-strength–pH plane depicting three regions: a clustered region, a suspended regime, and a repulsive structure. From [76]

controls the surface charge density which, in turn, affects the electrostatic interactions between the colloids. Increasing the ionic strength, experimentally achieved by "adding salt," decreases the screening length $1/\kappa$, so that the attractive forces become more important; the particles start forming clusters. Three different states are observed (1) a clustered regime, where particles aggregate when van der Waals attractions dominate, (2) a suspended regime where particles are distributed homogeneously and can move freely – corresponding to a stable suspension favored when electrostatic repulsion prevents clustering but is not strong enough to induce order. At very strong Coulomb repulsion the repulsive regime (3) occurs, where the mobility of the particles is restricted, and particles arrange in local order which maximizes nearest neighbor distances.

The location of the phase boundaries in Fig. 8 depends on the shear rate. In the clustered phase, shear leads to a breakup of clusters, and for the shear rate $\dot{\gamma} = 1,000\,\mathrm{s}^{-1}$, there are many small clusters which behave like single particles. In the regime where the particles are slightly clustered, or suspended, shear thinning is observed. Shear thinning is more pronounced in the slightly clustered state, because shear tends to reduce cluster size. Reasonable agreement with experiments was achieved, and discrepancies were attributed to polydispersity and the manner in which lubrication forces were approximated, as well as uncertainties how the pH and ionic strength enter the model force parameters.

In the simulations of Hecht et al. [16], the simple collisional coupling procedure described in Sect. 7.1 was used. This means that the colloids were treated as point particles, and solvent particles could flow right through them. Hydrodynamic interactions were therefore only resolved in an average sense, which is acceptable for studies of the general properties of an ensemble of many colloids. The heat from viscous heating was removed using the stochastic thermostat described in Sect. 2.3.

Various methods for modeling no-slip boundary conditions at colloid surfaces –
such as the thermal wall coupling described in Sect. 7.2 – were systematically investigated in [79]. No-slip boundary conditions are important, since colloids are
typically not completely spherical or smooth, and the solvent molecules also transfer
angular momentum to the colloid. Using the SRD algorithm without angular momentum conservation, it was found that the rotational friction coefficient was larger
than predicted by Enskog theory when the ghost-particle coupling was used [82]. On
the other hand, in a detailed study of the translational and rotational velocity autocorrelation function of a sphere coupled to the solvent by the thermal-wall boundary
condition, quantitative agreement with Enskog theory was observed at short times,
and with mode-coupling theory at long times. However, it was also noticed that
for small particles, the Enskog and hydrodynamic contributions to the friction coefficients were not clearly separated. Specifically, mapping the system to a density
matched colloid in water, it appeared that the Enskog and the hydrodynamic contributions are equal at a particle radius of 6 nm for translation and 35.4 nm for rotation;
even for a particle radius of 100 nm, the Enskog contribution to the friction is still
of order 30% and cannot be ignored.

In order to clarify the detailed character of the hydrodynamic interactions between colloids in SRD, Lee and Kapral [103] numerically evaluated the fixed-
particle friction tensor for two nano-spheres embedded in an SRD solvent. They
found that for intercolloidal spacings less than $1.2\,d$, where d is the colloid diameter, the measured friction coefficients start to deviate from the expected theoretical
curve. The reader is referred to the review by Kapral [30] for more details.

10.2 Polymer Dynamics

The dynamical behavior of macromolecules in solution is strongly affected or even
dominated by hydrodynamic interactions [6, 104, 105]. From a theoretical point of
view, scaling relations predicted by the Zimm model for, e.g., the dependencies of
dynamical quantities on the length of the polymer are, in general, accepted and confirmed [106]. Recent advances in experimental single-molecule techniques provide
insight into the dynamics of individual polymers, and raise the need for a quantitative theoretical description in order to determine molecular parameters such as
diffusion coefficients and relaxation times. Mesoscale hydrodynamic simulations
can be used to verify the validity of theoretical models. Even more, such simulations are especially valuable when analytical methods fail, as for more complicated
molecules such as polymer brushes, stars, ultrasoft colloids, or semidilute solutions,
where hydrodynamic interactions are screened to a certain degree. Here, mesoscale
simulations still provide a full characterization of the polymer dynamics.

We will focus on the dynamics of polymer chains in dilute solution. In order
to compare simulation results with theory – in particular the Zimm approach [6,
107] – and scaling predictions, we address the dynamics of Gaussian as well as self-
avoiding polymers.

10.2.1 Simulation Method and Model

Polymer molecules are composed of a large number of equal repeat units called monomers. To account for the generic features of polymers, such as their conformational freedom, no detailed modeling of the basic units is necessary. A coarse-grained description often suffices, where several monomers are comprised in an effective particle. Adopting such an approach, a polymer chain is introduced into the MPC solvent by adding N_m point particles, each of mass m_m, which are connected linearly by bonds. Two different models are considered, a Gaussian polymer and a polymer with excluded-volume (EV) interactions. Correspondingly, the following potentials are applied:

(1) *Gaussian chain:* The monomers, with the positions $\mathbf{r}_i$ ($i = 1, \ldots, N_m$), are connected by the harmonic potential

$$U_G = \frac{3k_B T}{2b^2} \sum_{i=1}^{N_m-1} (\mathbf{r}_{i+1} - \mathbf{r}_i)^2, \tag{81}$$

with zero mean bond length, and b the root-mean-square bond length. Here, the various monomers freely penetrated each other. This simplification allows for an analytical treatment of the chain dynamics as in the Zimm model [6, 107].

(2) *Excluded-volume chain:* The monomers are connected by the harmonic potential

$$U_B = \frac{\kappa}{2} \sum_{i=1}^{N_m-1} (|\mathbf{r}_{i+1} - \mathbf{r}_i| - b)^2, \tag{82}$$

with mean bond length b. The force constant κ is chosen such that the fluctuations of the bond lengths are on the order of a percent of the mean bond length. In addition, non-bonded monomers interact via the repulsive, truncated Lennard-Jones potential

$$U_{LJ} = \begin{cases} 4\epsilon \left[\left(\frac{\sigma}{r}\right)^{12} - \left(\frac{\sigma}{r}\right)^6 \right] + \epsilon, & r < 2^{1/6}\sigma, \\ 0, & \text{otherwise.} \end{cases} \tag{83}$$

The excluded volume leads to swelling of the polymer structure compared to a Gaussian chain, which is difficult to fully account for in analytical calculations [73].

The dynamics of the chain monomers is determined by Newtons' equations of motion between the collisions with the solvent. These equations are integrated using the velocity Verlet algorithm with the time step Δt_p. The latter is typically smaller than the collision time Δt. The monomer–solvent interaction is taken into account by inclusion of the monomer of mass $m_m = \rho m$ in the collision step [68, 73], compare Sect. 7.1. Alternatively, a Lennard-Jones potential can be used to account for the monomer–MPC particle interaction, where a MPC particle is of zero interaction range [19, 108].

We scale length and time according to $\hat{x} = x/a$ and $\hat{t} = t\sqrt{k_B T/ma^2}$, which corresponds to the choice $k_B T = 1$, $m = 1$, and $a = 1$. The mean free path of a fluid particle $\Delta t \sqrt{k_B T/m}$ is then given by $\lambda = \Delta \hat{t}$. In addition, we set $b = a$, $\sigma = a$, and $\epsilon/k_B T = 1$.

The equilibrium properties of a polymer are not affected by hydrodynamic interactions. Indeed, the results for various equilibrium quantities – such as the radius of gyration – of MPC simulations are in excellent agreement with the results of molecular dynamics of Monte Carlo simulations without explicit solvent [73].

Simulations of Gaussian chains, i.e., polymers with the bond potential (81), can be compared with analytical calculations based on the Zimm approach [6, 107]. Note, however, that the simulations are *not* performed in the Zimm model. The Zimm approach relies on the preaveraging approximation of hydrodynamic interactions, whereas the simulations take into account the configurational dependence of the hydrodynamic interactions, and therefore hydrodynamic fluctuations. Hence, the comparison can serve as a test of the validity of the approximations employed in the Zimm approach.

The Zimm model rests upon the Langevin equation for over-damped motion of the monomers, i.e., it applies for times larger than the Brownian time scale $\tau_B \gg m_m/\zeta$, where ζ is Stokes' friction coefficient [12]. On such time scales, velocity correlation functions have decayed to zero and the monomer momenta are in equilibrium with the solvent. Moreover, hydrodynamic interactions between the various parts of the polymer are assumed to propagate instantaneously. This is not the case in our simulations. First of all, the monomer inertia term is taken into account, which implies non-zero velocity autocorrelation functions. Secondly, the hydrodynamic interactions build up gradually. The center-of-mass velocity autocorrelation function displayed in Fig. 9 reflects these aspects. The correlation function exhibits a long-time tail, which decays as $\langle \mathbf{v}_{cm}(t) \mathbf{v}_{cm}(0) \rangle \sim t^{-3/2}$ on larger time scales. The algebraic decay is associated with a coupling between the motion of the polymer and the hydrodynamic modes of the fluid [99, 109, 110]. A scaling of time with

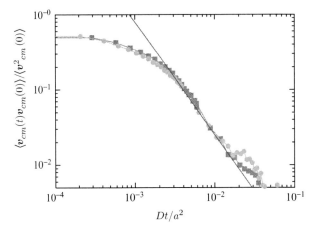

Fig. 9 Center-of-mass velocity autocorrelation functions for Gaussian polymers of length $N_m = 20$, $N_m = 40$, and $\lambda = 0.1$ as a function of Dt. The *solid line* is proportional to $(Dt)^{-3/2}$. From [73]

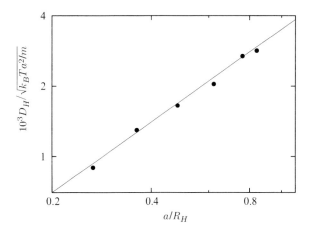

Fig. 10 Dependence of the hydrodynamic part of the diffusion coefficient, $D_H = D - D_0/N_m$, on the hydrodynamic radius for Gaussian chains of lengths $N_m = 5, 10, 20, 40, 80,$ and 160 (*right to left*). The mean free path is $\lambda = 0.1$. From [73]

the diffusion coefficient D shows that the correlation function is a universal function of Dt. This is in agreement with results of DPD simulations of dilute polymer systems [92].

The polymer center-of-mass diffusion coefficient follows either via the GK relation from the velocity autocorrelation function or by the Einstein relation from the center-of-mass mean square displacement. According to the Kirkwood formula [104, 105, 111]

$$D^{(K)} = \frac{D_0}{N_m} + \frac{k_B T}{6\pi\eta} \frac{1}{R_H},$$ (84)

where the hydrodynamic radius R_H is defined as

$$\frac{1}{R_H} = \frac{1}{N_m^2} \left\langle \sum_{i=1}^{N_m} \sum_{j=1}^{N_m}{}' \frac{1}{|\mathbf{r}_i - \mathbf{r}_j|} \right\rangle$$ (85)

and the prime indicates that the term with $j = i$ has to be left out in the summation. The diffusion coefficient is composed of the local friction contribution D_0/N_m, where D_0 is the diffusion coefficient of a single monomer in the same solvent, and the hydrodynamic contribution.

Simulation results for the hydrodynamic contribution, $D_H = D - D_0/N_m$, to the diffusion coefficient are plotted in Fig. 10 as a function of the hydrodynamic radius (85). In the limit $N_m \gg 1$, the diffusion coefficient D is dominated by the hydrodynamic contribution D_H, since $D_H \sim N_m^{-1/2}$. For shorter chains, D_0/N_m cannot be neglected, and therefore has to be subtracted in order to extract the scaling behavior of D_H. The hydrodynamic part of the diffusion coefficient D_H exhibits the dependence predicted by the Kirkwood formula and the Zimm theory, i.e., $D_H \sim 1/R_H$. The finite-size corrections to D show a dependence $D = D_\infty - \text{const.}/L$ on the size

L of a periodic system, in agreement with previous studies [68, 91, 112]. Simulations for various system sizes for polymers of lengths $N_m = 10, 20$, and 40 allow an extrapolation to infinite system size, which yields $D_0/\sqrt{k_B T a^2/m} \approx 1.7 \times 10^{-2}$, in good agreement with the diffusion coefficient of a monomer in the same solvent. The values of D_∞ are about 30% larger than the finite-system-size values presented in Fig. 10. Similarly the diffusion coefficient for a polymer chain with excluded volume interactions displays the dependence $D_H \sim 1/R_H$ [73].

The Kirkwood formula neglects hydrodynamic fluctuations and is thus identical with the preaveraging result of the Zimm approach. When only the hydrodynamic part is considered, the Zimm model yields the diffusion coefficient

$$D_Z = 0.192 \frac{k_B T}{b\eta \sqrt{N_m}}. \tag{86}$$

MPC simulations for polymers of length $N_m = 40$ yield $D_Z/\sqrt{k_B T a^2/m} = 0.003$. This value agrees with the numerical value for an infinite system, $D_H/\sqrt{k_B T a^2/m} = 0.0027$, within 10%. The MPC simulations yield a diffusion coefficient smaller than $D^{(K)}$, in agreement with previous studies presented in [6, 111, 113]. Note that the experimental values are also smaller by about 15% than those predicted by the Zimm approach [6, 114, 115].

To further characterize the internal dynamics of the molecular chain, a mode analysis in terms of the eigenfunctions of the discrete Rouse model [6, 116] has been performed. The mode amplitudes χ_p are calculated according to

$$\chi_p = \sqrt{\frac{2}{N_m}} \sum_{i=1}^{N_m} \mathbf{r}_i \cos\left[\frac{p\pi}{N_m}\left(i - \frac{1}{2}\right)\right], \quad p = 1, \ldots, N_m. \tag{87}$$

Because of hydrodynamic interactions, Rouse modes are no longer eigenfunctions of the chain molecule. However, within the Zimm theory, they are reasonable approximations and the autocorrelation functions of the mode amplitudes decay exponentially, i.e.,

$$\left\langle \chi_p(t)\chi_p(0) \right\rangle = \left\langle \chi_p^2 \right\rangle \exp\left(-t/\tau_p\right). \tag{88}$$

For the Rouse model, the relaxation times τ_p depend on chain length and mode number according to $\tau_p \sim 1/\sin^2\left(p\pi/2N_m\right)$, whereas for the Zimm model the dependence

$$\tau_p \sim (p/N_m)^{1/2}/\sin^2\left(p\pi/2N_m\right) \tag{89}$$

is obtained. The extra contribution $\sqrt{p/N_m}$ follows from the eigenfunction representation of the preaveraged hydrodynamic tensor, under the assumption that its off-diagonal elements do not significantly contribute to the relaxation behavior.

In Fig. 11, the autocorrelation functions for the mode amplitudes are shown for the mean free path $\lambda = 0.1$. Within the accuracy of the simulations, the correlation functions decay exponentially and exhibit the scaling behavior predicted by the Zimm model. Hence, for the small mean free path, hydrodynamic interactions are

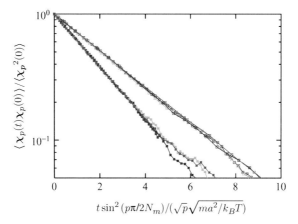

Fig. 11 Correlation functions of the Rouse-mode amplitudes for the modes $p = 1 - 4$ of Gaussian polymers. The chain lengths are $N_m = 20$ (*right*) and $N_m = 40$ (*left*). From [73]

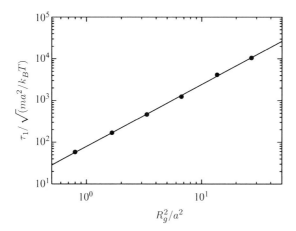

Fig. 12 Dependence of the longest relaxation time τ_1 on the radius of gyration for Gaussian chains of the lengths given in Fig. 10. From [73]

taken into account correctly. This is no longer true for the large mean free path, $\lambda = 2$. In this case, a scaling behavior between that predicted by the Rouse and Zimm models is observed. This implies that hydrodynamic interactions are present, but are not fully developed or are small compared to the local friction of the monomers. We obtain pure Rouse behavior for a system without solvent by simply rotating the velocities of the individual monomers [73].

The dependence of the longest relaxation time on the radius of gyration is displayed in Fig. 12 for $\lambda = 0.1$. The scaling behavior $\tau_1 \sim R_G^3$ is in very good agreement with the predictions of the Zimm theory. We even find almost quantitative agreement; the relaxation time of the $p = 1$ mode of our simulations is approximately 30% larger than the Zimm value [6].

The scaling behavior of *equilibrium* properties of single polymers with excluded-volume interactions has been studied extensively [6,117–120]. It has been found that even very short chains already follow the scaling behavior expected for much longer chains. In particular, the radius of gyration increases like $R_G \sim N_m^\nu$ with the number of monomers, and the static structure factor $S(\mathbf{q})$ exhibits a scaling regime for $2\pi/R_G \ll q \ll 2\pi/\sigma$, with a $q^{-1/\nu}$ decay as a function of the scattering vector q and the exponent $\nu \approx 0.6$. For the interaction potentials (82), (83) with the parameters $b = \sigma = a$, $\epsilon/k_B T = 1$, the exponent $\nu \approx 0.62$ is obtained from the chain-length dependence of the radius of gyration, the mean square end-to-end distance, as well as the q–dependence of the static structure factor [73].

An analysis of the intramolecular dynamics in terms of the Rouse modes yields non-exponentially decaying autocorrelation functions of the mode amplitudes. At very short times, a fast decay is found, which turns into a slower exponential decay which is well fitted by $A_p \exp(-t/\tau_p)$, see Fig. 13. Within the accuracy of these calculations, the correlation functions exhibit universal behavior. Zimm theory predicts the dependence $\tau_p \sim p^{-3\nu}$ for the relaxation times on the *mode number* for polymers with excluded-volume interactions [6]. With $\nu = 0.62$, the exponent α for the polymer of length $N_m = 40$ is found to be in excellent agreement with the theoretical prediction. The exponent for the polymers with $N_m = 20$ is slightly larger.

Zimm theory predicts that the dynamic structure factor, which is defined by

$$S(\mathbf{q},t) = \frac{1}{N_m} \sum_{i=1}^{N_m} \sum_{j=1}^{N_m} \left\langle \exp\left(i\mathbf{q}[\mathbf{r}_i(t) - \mathbf{r}_j(0)]\right) \right\rangle, \tag{90}$$

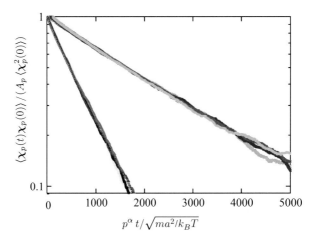

Fig. 13 Correlation functions of the Rouse-mode amplitudes for various modes as a function of the scaled time tp^α for polymers with excluded volume interactions. The chain lengths are $N_m = 20$ (*left*) and $N_m = 40$ (*right*). The calculated correlations were fitted by $A_p \exp(-t/\tau_p)$ and have been divided by A_p. The scaling exponents of the mode numbers are $\alpha = 1.93$ ($N_m = 20$) and $\alpha = 1.85$ ($N_m = 40$), respectively. From [73]

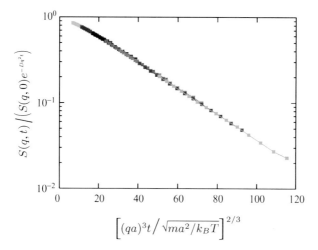

Fig. 14 Normalized dynamic structure factor $S(\mathbf{q},t)/(S(\mathbf{q},0)\exp\left(-Dq^2t\right))$ of polymers with excluded volume interactions for $N_m = 40$ and various q-values in the range $0.7 < qa < 2$ as a function of $q^2t^{2/3}$. From [73]

scales as [6]

$$S(\mathbf{q},t) = S(\mathbf{q},0)f(q^\alpha t) \tag{91}$$

with $\alpha = 3$ for $qR_G \gg 1$, independent of the solvent conditions (Θ or good solvent). To extract the scaling relation for the intramolecular dynamics, which corresponds to the prediction (91), we resort to the following considerations. As is well known, the dynamic structure factor for a Gaussian distribution of the differences $\mathbf{r}_i(t) - \mathbf{r}_j(0)$ and a linear equation of motion is given by [6, 121]

$$S(\mathbf{q},t) = S(\mathbf{q},0)e^{-Dq^2t}\frac{1}{N_m}\sum_{i=1}^{N_m}\sum_{j=1}^{N_m}\exp\left(-q^2\left\langle(\mathbf{r}_i'(t) - \mathbf{r}_j'(0))^2\right\rangle/6\right), \tag{92}$$

where Dq^2t accounts for the center-of-mass dynamics and $\mathbf{r}_i'$ denotes the position of monomer i in the center-of-mass reference frame. Therefore, in order to obtain the dynamics in the center-of-mass reference frame, we plot $S(\mathbf{q},t)/(S(\mathbf{q},0)\exp\left(-Dq^2t\right))$. The simulation results for the polymer of length $N_m = 40$, shown in Fig. 14, confirm the predicted scaling behavior. Thus, MPC–MD hybrid simulations are very well suited to study the dynamics of even short polymers in dilute solution.

As mentioned above, the structure of a polymer depends on the nature of the solvent. In good solvent excluded volume interactions lead to expanded conformations and under bad solvent conditions the polymer forms a dense coil. In a number of simulations the influence of hydrodynamic interactions on the transition from an extended to a collapsed state has been studied, when the solvent quality is abruptly changed. Both, molecular dynamics simulations with an explicit solvent [122] as well as MPC simulations [108, 123, 124] yield significantly different dynamics in the presence of hydrodynamic interactions. Specifically, the collapse is faster, with a much weaker dependence of the characteristic time on polymer length [108, 124], and the folding path is altered.

Similarly, a strong influence of hydrodynamic interactions has been found on the polymer translocation dynamics through a small hole in a wall [125] or in polymer packing in a virus capsid [126, 127]. Cooperative backflow effects lead to a rather sharp distribution of translocation times with a peak at relatively short times. The fluid flow field, which is created as a monomer moves through the hole, guides following monomers to move in the same direction.

10.3 Polymers in Flow Fields

Simulations of an MPC fluid confined between surfaces and exposed to a constant external force yield the expected parabolic velocity profile for appropriate boundary conditions [31, 36, 128]. The ability of MPC to account for the flow behavior of mesoscale objects, such as polymers, under non-equilibrium conditions has been demonstrated for a number of systems. Rod-like colloids in shear flow exhibit flow induced alignment [72]. The various diagonal components of the radius of gyration tensor exhibit qualitatively and quantitatively a different behavior. Because of the orientation, the component in the flow direction increases with increasing Peclet number larger than unity and saturates at large shear rates because of finite size effects. The transverse components decrease with shear rate, where the component in the gradient direction is reduced to a greater extent. The rod rotational velocity in the shear plane shows two distinct regimes. For Peclet numbers much smaller than unity, the rotational velocity increases linearly with the shear rate, because the system is isotropic. At Peclet numbers much larger than unity, the shear-induced anisotropies lead to a slower increase of the rotational velocity with the shear rate [72].

The simulations of a tethered polymer in a Poiseuille flow [74] yield a series of morphological transitions from sphere to deformed sphere to trumpet to stem and flower to rod, similar to theoretically predicted structures [129–131]. The crossovers between the various regimes occur at flow rates close to the theoretical estimates for a similar system. Moreover, the simulations in [74] show that backflow effects lead to an effective increase in viscosity, which is attributed to the fluctuations of the free polymer end rather than its shape.

The conformational, structural, and transport properties of free flexible polymers in microchannel flow have been studied in [128, 132] by hybrid MPC–MD simulations. These simulations confirm the cross streamline migration of the molecules as previously observed in [133–138]. In addition, various other polymer properties are addressed in [132].

All these hybrid simulations confirm that MPC is an excellent method to study the non-equilibrium behavior of polymers in flow fields. In the next section, we will provide a more detailed example for a more complicated object, namely an ultrasoft colloid in shear flow.

10.4 Ultra-Soft Colloids in Shear Flow

Star polymers present a special macromolecular architecture, in which several linear polymers of identical length are linked together by one of their ends at a common center. This structure is particularly interesting because it allows for an almost continuous change of properties from that of a flexible linear polymer to a spherical colloidal particle with very soft interactions. Star polymers are therefore also often called ultrasoft colloids. The properties of star polymers in and close to equilibrium have been studied intensively, both theoretically [139,140] and experimentally [141]. A star polymer is a ultrasoft colloid, where the core extension is very small compared to the length of an arm. By anchoring polymers on the surface of a hard colloid, the softness can continuously be changed from ultrasoft to hard by increasing the ratio between the core and shell radius at the expense of the thickness of the soft polymer corona. Moreover, star polymers have certain features in common with vesicles and droplets. Although their shell can be softer than that of the other objects, the dense packing of the monomers will lead to a cooperative dynamical behavior resembling that of vesicles or droplets [142].

Vesicles and droplets encompass fluid which is not exchanged with the surrounding. In contrast, for star-like molecules fluid is free to penetrate into the molecule and internal fluid is exchanged with the surrounding in the course of time. This intimate coupling of the star-polymer dynamics and the fluid flow leads to a strong modification of the flow behavior at and next to the ultrasoft colloid particularly in non-equilibrium systems.

In the following, we will discuss a few aspects in the behavior of star polymers in shear flow as a function of arm number f, arm length L_f, and shear rate $\dot{\gamma}$. The polymer model is the same as described in Sect. 10.2.1, where the chain connectivity is determined by the bond potential (82) and the excluded-volume interaction is described by the Lennard-Jones potential (83). A star polymer of functionality f is modeled as f linear polymer chains of L_f monomers each, with one of their ends linked to a central particle. Linear polymer molecules are a special case of star polymers with functionality $f = 2$. Lees–Edwards boundary conditions [42] are employed in order to impose a linear velocity profile $(v_x, v_y, v_z) = (\dot{\gamma} r_y, 0, 0)$ in the fluid in the absence of a polymer. For small shear rates, the conformations of star polymers remain essentially unchanged compared to the equilibrium state. Only when the shear rate exceeds a characteristic value, a structural anisotropy as well as an alignment is induced by the flow (cf. Fig. 15). The shear rate dependent quantities are typically presented in terms of the Weissenberg number $\mathrm{Wi} = \dot{\gamma}\tau$ rather than the shear rate itself, where τ is the longest characteristic relaxation time of the considered system. For the star polymers, the best data collapse for stars of various arm lengths is found when the relaxation time $\tau = \eta b^3 L_f^2 / k_B T$ is used [142]. Remarkably, there is essentially no dependence on the functionality. Within the range of investigated star sizes, this relaxation time has to be considered as consistent with the prediction for the blob model of [143], where $\tau \sim L_f^{1.8} f^{0.1}$ for the Flory exponent $\nu = 0.6$.

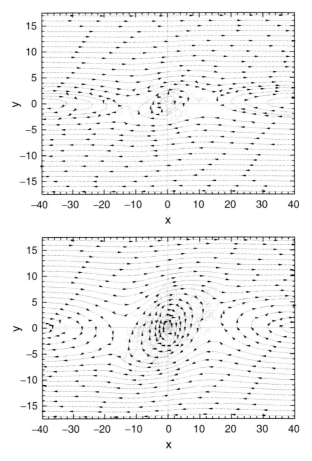

Fig. 15 Fluid flow lines in the flow-gradient plane of the star polymer's center of mass reference frame for $f = 10$ (*top*) and $f = 50$ (*bottom*) arms, both with $L_f = 20$ monomers per arm and an applied shear field with Wi $= \dot{\gamma}\tau = 22$. From [25]

In Fig. 15, typical star conformations are shown which indicate the alignment and induced anisotropy in the flow. Moreover, the figure reveals the intimate coupling of the polymer dynamics and the emerging fluid flow field. In the region, where the fluid coexists with the star polymer, the externally imposed flow field is strongly screened and the fluid velocity is no longer aligned with the shear flow direction, but rotates around the polymer center of mass. The fluid stream lines are calculated by integration of the coarse-grained fluid velocity field. Outside the region covered by the star polymer, the fluid adapts to the central rotation by generating two counter-rotating vortices, and correspondingly two *stagnation points* of vanishing fluid velocity [25].

The fluid flow in the vicinity of the star polymer is distinctively different from that of a sphere but resembles the flow around an ellipsoid [144]. In contrast to the latter, the fluid penetrates into the area covered by the star polymer. While the fluid in the core of the star rotates together with the polymer, the fluid in the corona follows the external flow to a certain extent.

A convenient quantity to characterize the structural properties and alignment of polymers in flow is the average gyration tensor, which is defined as

$$G_{\alpha\beta}(\dot{\gamma}) = \frac{1}{N_m} \sum_{i=1}^{N_m} \langle r'_{i,\alpha} r'_{i,\beta} \rangle , \qquad (93)$$

where $N_m = fL_f + 1$ is the total number of monomers, $r'_{i,\alpha}$ is the position of monomer i relative to the polymer center of mass, and $\alpha \in \{x, y, z\}$. The average gyration tensor is directly accessible in scattering experiments. Its diagonal components $G_{\alpha\alpha}(\dot{\gamma})$ are the squared radii of gyration of the star polymer along the axes of the reference frame. In the absence of flow, scaling considerations predict [139] $G_{xx}(0) = G_{yy}(0) = G_{zz}(0) = R_G^2(0)/3 \sim L_f^{2\nu} f^{1-\nu}$.

The diagonal components $G_{\alpha\alpha}$ of the average gyration tensor are shown in Fig. 16 as a function of the Weissenberg number for various functionalities and arm lengths. We find that the extension of a star increases with increasing shear rate in the shear direction (x), decreases in the gradient direction (y), and is almost independent of Wi in the vorticity direction (z). The deviation from spherical symmetry exhibits a Wi^2 power-law dependence for small shear rates for all functionalities. A similar behavior has been found for rod-like colloids [72] (due to the increasing alignment with the flow direction) and for linear polymers [6]. However, for stars of not too small functionality, a new scaling regime appears, where the deformation seems to scale *linearly* with the Weissenberg number. For large shear rates,

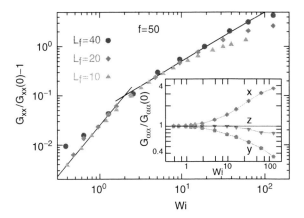

Fig. 16 Normalized component G_{xx} of the average gyration tensor as a function of the Weissenberg number Wi, for star polymers of $f = 50$ arms and the arm lengths $L_f = 10$, 20, and 40 monomers. Power-law behaviors with quadratic and linear dependencies on Wi are indicated by lines. The *inset* shows all three diagonal components of the gyration tensor (for $L_f = 20$). From [142]

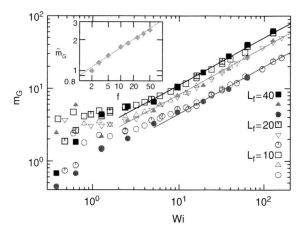

Fig. 17 Orientational resistance m_G as a function of the Weissenberg number Wi for star polymers with functionalities $f = 2$ (*circles*), $f = 15$ (*triangles*), and $f = 50$ (*squares*), and different arm lengths indicated in the figure. *Lines* correspond to the power law $m_G = \tilde{m}_G(f)\text{Wi}^{0.65}$. The *inset* shows that the amplitude also follows a power-law behavior with $\tilde{m}_G(f) \sim f^{0.27}$. From [142]

finite-size effects appear due to the finite monomer number. These effects emerge when the arms are nearly stretched, and therefore occur at higher Weissenberg numbers for larger arm lengths.

The average flow alignment of a (star) polymer can be characterized by the orientation angle χ_G, which is the angle between the eigenvector of the gyration tensor with the largest eigenvalue and the flow direction. It follows straightforwardly [87] from the simulation data via

$$\tan(2\chi_G) = 2G_{xy}/(G_{xx} - G_{yy}) \equiv m_G/\text{Wi}, \tag{94}$$

where the right-hand-side of the equation defines the orientation resistance parameter m_G [145]. It has been shown for several systems including rod-like colloids and linear polymers without self-avoidance [6] that close to equilibrium $G_{xy} \sim \dot{\gamma}$ and $(G_{xx} - G_{yy}) \sim \dot{\gamma}^2$, so that m_G is independent of Wi. Our results for the orientation resistance are presented in Fig. 17 for various functionalities f and arm lengths L_f. Data for different L_f collapse onto universal curves, which approach a plateau for small shear rates, as expected. For larger shear rates, Wi $\gg 1$, a power-law behavior [142]

$$m_G(\text{Wi}) \sim f^\alpha \, \text{Wi}^\mu, \tag{95}$$

is obtained with respect to the Weissenberg number and the functionality, where $\alpha = 0.27 \pm 0.02$ and $\mu = 0.65 \pm 0.05$. For self-avoiding linear polymers, a somewhat smaller exponent $\mu = 0.54 \pm 0.03$ was obtained in [87, 146], whereas theoretical calculations predict $\sim\text{Wi}^{2/3}$ in the limit of large Weissenberg numbers [147].

The data for the average orientation and deformation of a star polymer described so far seem to indicate that the properties vary smoothly and monotonically from lin-

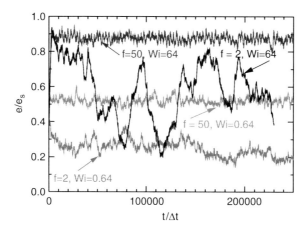

Fig. 18 Temporal evolution of the largest intramolecular distance $e = \max_{ij} |\mathbf{r}_i - \mathbf{r}_j|$ of a linear polymer and a star polymer with 50 arms, for the Weissenberg numbers $Wi = 0.64$ and $Wi = 64$. In both cases, $L_f = 40$. The time t is measured in units of the collision time δt. e_s corresponds to the fully stretched arms. From [142]

ear polymers to star polymers of high functionality. However, this picture changes when the *dynamical behavior* is considered. It is well known by now that linear polymers show a tumbling motion in flow, with alternating collapsed and stretched configurations during each cycle [146, 148–150]. For large Weissenberg numbers, this leads to very large fluctuations of the largest intramolecular distance of a linear polymer with time, as demonstrated experimentally in [146, 150], and reproduced in the MPC simulations [142], see Fig. 18. A similar behavior is found for $f = 3$. However, for $f > 5$, a quantitatively different behavior is observed as displayed in Fig. 19. Now, the fluctuations of the largest intramolecular distance are much smaller and *decrease* with increasing Weissenberg number as shown in Fig. 18, and the dynamics resembles much more the continuous tank-treading motion of fluid droplets and capsules. The shape and orientation of such stars depends very little on time, while the whole object is rotating. On the other hand, a single, selected arm resembles qualitatively the behavior of a linear polymer – it also collapses and stretches during the tank-treading motion. The successive snapshots of Fig. 20 illustrate the tank-treading motion. Following the top left red polymer in the top left image, we see that the extended polymer collapses in the course of time, moves to the right and stretches again. In parallel, other polymers exhibit a similar behavior on the bottom side. Moreover, the images show that the orientation of a star hardly changes in the course of time.

The rotational dynamics of a star polymer can be characterized quantitatively by calculating the rotation frequency

$$\omega_\alpha = \sum_{\beta=x}^{z} \langle \Theta_{\alpha\beta}^{-1} L_\beta \rangle \tag{96}$$

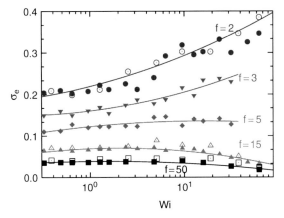

Fig. 19 Widths of the distribution functions of the largest intramolecular distances, $\sigma_e = (\langle e^2 \rangle - \langle e \rangle^2)/\langle e \rangle^2$, of a linear polymer and star polymers with up to 50 arms as a function of the Weissenberg number. From [142]

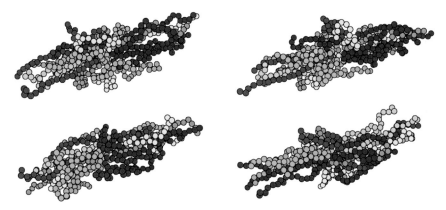

Fig. 20 Successive snapshots of a star polymer of functionality $f = 25$ and arm length $L_f = 20$, which illustrate the tank-treading motion

of a star, where

$$\Theta_{\alpha\beta} = \sum_{i=1}^{N_m} \left[\mathbf{r}_i'^2 \delta_{\alpha\beta} - r_{i,\alpha}' r_{i,\beta}' \right] \tag{97}$$

is the instantaneous moment-of-inertia tensor and L_β is the instantaneous angular momentum. Since the rotation frequency for all kinds of soft objects – such as rods, linear polymers, droplets and capsules – depends linearly on $\dot{\gamma}$ for small shear rates, the reduced rotation frequency $\omega/\dot{\gamma}$ is shown in Fig. 21 as a function of the Weissenberg number. The data approach $\omega/\dot{\gamma} = 1/2$ for small Wi, as expected [87, 151]. For larger shear rates, the reduced frequency decreases due to the deformation and alignment of the polymers in the flow field. With increasing arm number, the decrease of $\omega/\dot{\gamma}$ at a given Weissenberg number becomes smaller, since the deviation

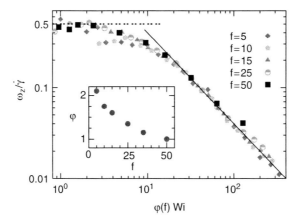

Fig. 21 Scaled rotation frequencies as a function of a rescaled Weissenberg number for various functionalities. *Dashed and full lines* correspond to $\omega/\dot\gamma = 1/2$ for small Wi, and $\omega/\dot\gamma \sim 1/\text{Wi}$ for large Wi, respectively. The *inset* shows the dependence of the rescaling factor φ on the functionality. From [142]

from the spherical shape decreases. Remarkably, the frequency curves for all stars with $f > 5$ are found to collapse onto a universal scaling function when $\omega/\dot\gamma$ is plotted as a function of a rescaled Weissenberg number, see Fig. 21. For high shear rates, $\omega/\dot\gamma$ decays as Wi^{-1}, which implies that the rotation frequency becomes *independent* of $\dot\gamma$. A similar behavior has been observed for capsules at high shear rates [152].

The presented results show that star polymers in shear flow show a very rich structural and dynamical behavior. With increasing functionality, stars in flow change from linear-polymer-like to capsule-like behavior. These macromolecules are therefore interesting candidates to tune the viscoelastic properties of complex fluids.

11 Vesicles and Cells in Hydrodynamic Flows

11.1 Introduction

The flow behavior of fluid droplets, capsules, vesicles, and cells is of enormous importance in science and technology. For example, the coalescence and break-up of fluid droplets is essential for emulsion formation and stability. Capsules and vesicles are discussed and used as drug carriers. Red blood cells (RBC) flow in the blood stream, and may coagulate or be torn apart under unfavorable flow conditions. Red blood cells also have to squeeze through narrow capillaries to deliver their oxygen cargo. White blood cells in capillary flow adhere to, roll along and detach again from the walls of blood vessels under normal physiological conditions; in inflamed tissue, leukocyte rolling leads to firm adhesion and induces an immunological response.

These and many other applications have induced an intensive theoretical and simulation activity to understand and predict the behavior of such soft, deformable objects in flow. In fact, there are some general, qualitative properties in simple shear flow, which are shared by droplets, capsules, vesicles, and cells. When the internal viscosity is low and when they are highly deformable, a tank-treading motion is observed (in the case of droplets for not too high shear rates), where the shape and orientation are stationary, but particles localized at the interface or attached to the membrane orbit around the center of mass with a rotation axis in the vorticity direction. On the other hand, for high internal viscosity or small deformability, the whole object performs a tumbling motion, very much like a colloidal rod in shear flow. However, if we take a more careful look, then the behavior of droplets, capsules, vesicles, and cells is quite different. For example, droplets can break up easily at higher shear rates, because their shape is determined by the interfacial tension; fluid vesicles can deform much more easily then capsules, since their membrane has no shear elasticity; etc. We focus here on the behavior of fluid vesicles and red blood cells.

11.2 Modeling Membranes

11.2.1 Modeling Lipid-Bilayer Membranes

The modeling of lipid bilayer membranes depends very much on the length scale of interest. The structure of the bilayer itself or the embedding of membrane proteins in a bilayer are best studied with *atomistic models* of both lipid and water molecules. Molecular dynamics simulations of such models are restricted to about 10^3 lipid molecules. For larger system sizes, *coarse-grained models* are required [153–155]. Here, the hydrocarbon chains of lipid molecules are described by short polymer chains of Lennard-Jones particles, which have a repulsive interaction with the lipid head groups as well as with the water molecules, which are also modeled as single Lennard-Jones spheres. Very similar models, with Lennard-Jones interactions replaced by linear "soft" DPD potentials, have also been employed intensively [156–160]. For the investigation of shapes and thermal fluctuations of single- or multi-component membranes, the hydrodynamics of the solvent is irrelevant. In this case, it can be advantageous to use a *solvent-free membrane model*, in which the hydrophobic effect of the water molecules is replaced by an effective attraction among the hydrocarbon chains [161–164]. This approach is advantageous in the case of membranes in dilute solution, because it reduces the number of molecules – and thus the degrees of freedom to be simulated – by orders of magnitude. However, it should be noticed that the basic length scale of atomistic and coarse-grained or solvent-free models is still on the same order of magnitude.

In order to simulate larger systems, such as giant unilamellar vesicles (GUV) or red blood cells, which have a radius on the order of several micrometers, a different approach is required. It has been shown that in this limit the properties of lipid bilayer membranes are described very well by modeling the membrane as a two-dimensional manifold embedded in three-dimensional space, with the shape and fluctuations controlled by the curvature elasticity [165],

$$\mathcal{H} = \int dS\, 2\kappa H^2 , \tag{98}$$

where $H = (c_1 + c_2)/2$ is the mean curvature, with the local principal curvatures c_1 and c_2, and the integral is over the whole membrane area. To make the curvature elasticity amenable to computer simulations, it has to be discretized. This can be done either by using triangulated surfaces [166, 167], or by employing particles with properly designed interactions which favor the formation of self-assembled, nearly planar sheets [168, 169]. In the latter case, both scalar particles with isotropic multi-particle interactions (and a curvature energy obtained from a moving least-squares method) [169] as well as particles with an internal spin variable and anisotropic, multi-body forces [168] have been employed and investigated.

11.2.2 Dynamically Triangulated Surfaces

In a dynamically triangulated surface model [166, 167, 170–172] of vesicles and cells, the membrane is described by N_{mb} vertices which are connected by tethers to form a triangular network of spherical topology, see Fig. 22. The vertices have excluded volume and mass m_{mb}. Two vertices connected by a bond have an attractive interaction, which keeps their distance below a maximum separation ℓ_0. A short-range repulsive interaction among all vertices makes the network self-avoiding and prevents very short bond lengths. The curvature energy can be discretized in different ways [166, 173]. In particular, the discretization [173, 174]

$$U_{cv} = \frac{\kappa}{2} \sum_i \frac{1}{\sigma_i} \left\{ \sum_{j(i)} \frac{\sigma_{i,j} \mathbf{r}_{i,j}}{r_{i,j}} \right\}^2 \tag{99}$$

has been found to give reliable results in comparison with the continuum expression (98). Here, the sum over $j(i)$ is over the neighbors of a vertex i which are connected by tethers. The bond vector between the vertices i and j is $\mathbf{r}_{i,j}$, and $r_{i,j} = |\mathbf{r}_{i,j}|$. The length of a bond in the dual lattice is $\sigma_{i,j} = r_{i,j}[\cot(\theta_1) + \cot(\theta_2)]/2$, where the angles θ_1 and θ_2 are opposite to bond ij in the two triangles sharing this bond. Finally, $\sigma_i = 0.25 \sum_{j(i)} \sigma_{i,j} r_{i,j}$ is the area of the dual cell of vertex i.

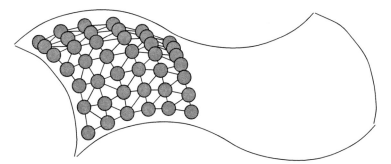

Fig. 22 Triangulated-network model of a fluctuating membrane. All vertices have short-range repulsive interactions symbolized by hard spheres. Bonds represent attractive interactions which imply a maximum separation ℓ_0 of connected vertices. From [175]

To model the fluidity of the membrane, tethers can be flipped between the two possible diagonals of two adjacent triangles. A number ψN_b of bond-flip attempts is performed with the Metropolis Monte Carlo method [173] at time intervals Δt_{BF}, where $N_b = 3(N_{mb} - 2)$ is the number of bonds in the network, and $0 < \psi < 1$ is a parameter of the model. Simulation results show that the vertices of a dynamically triangulated membrane show diffusion, i.e., the squared distance of two initially neighboring vertices increases linearly in time.

11.2.3 Vesicle Shapes

Since the solubility of lipids in water is very low, the number of lipid molecules in a membrane is essentially constant over typical experimental time scales. Also, the osmotic pressure generated by a small number of ions or macromolecules in solution, which cannot penetrate the lipid bilayer, keeps the internal volume essentially constant. The shape of fluid vesicles [176] is therefore determined by the competition of the curvature elasticity of the membrane, and the constraints of constant volume V and constant surface area S. In the simplest case of vanishing spontaneous curvature, the curvature elasticity is given by (98). In this case, the vesicle shape in the absence of thermal fluctuations depends on a single dimensionless parameter, the reduced volume $V^* = V/V_0$, where $V_0 = (4\pi/3)R_0^3$ and $R_0 = (S/4\pi)^{1/2}$ are the volume and radius of a sphere of the same surface area S, respectively. The calculated vesicle shapes are shown in Fig. 23. There are three phases. For reduced volumes not too far from the sphere, elongated prolate shapes are stable. In a small range of reduced volumes of $V^* \in [0.592, 0.651]$, oblate discocyte shapes have the lowest curvature energy. Finally, at very low reduced volumes, cup-like stomatocyte shapes are found.

These shapes are very well reproduced in simulations with dynamically triangulated surfaces [177–180].

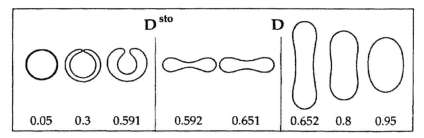

Fig. 23 Shapes of fluid vesicles as a function of the reduced volume V^*. D and D^{sto} denote the discontinuous prolate–oblate and oblate–stomatocyte transitions, respectively. All shapes display rotational symmetry with respect to the vertical axis. From [176]

11.2.4   Modeling Red Blood Cells

Red blood cells have a biconcave disc shape, which can hardly be distinguished from the discocyte shape of fluid vesicles with reduced volume $V^* \simeq 0.6$, compare Fig. 23. However, the membrane of red blood cells is more complex, since a spectrin network is attached to the plasma membrane [181], which helps to retain the integrity of the cell in strong shear gradients or capillary flow. Because of the spectrin network, the red blood cell membrane has a non-zero shear modulus μ.

The bending rigidity κ of RBCs has been measured by micropipette aspiration [182] and atomic force microscopy [183] to be approximately $\kappa = 50 k_B T$. The shear modulus of the composite membrane, which is induced by the spectrin network, has been determined by several techniques; it is found to be $\mu = 2 \times 10^{-6}\,\text{N}\,\text{m}^{-1}$ from optical tweezers manipulation [184], while the value $\mu = 6 \times 10^{-6}\,\text{N}\,\text{m}^{-1}$ is obtained from micropipette aspiration [182]. Thus, the dimensionless ratio $\mu R_0^2/\kappa \simeq 100$, which implies that bending and stretching energies are roughly of equal importance.

Theoretically, the shapes of RBCs in the absence of flow have been calculated very successfully on the basis of a mechanical model of membranes, which includes both curvature and shear elasticity [185,186]. In particular, it has been shown recently that the full stomatocyte–discocyte–echinocyte sequence of RBCs can be reproduced by this model [186].

The composite membrane of a red blood cell, consisting of the lipid bilayer and the spectrin network, can be modeled as a composite network, which consists of a dynamically triangulated surface as in the case of fluid vesicles, coupled to an additional network of harmonic springs with fixed connectivity (no bond-flip) [185, 187]. Ideally, the bond length of the elastic network is larger than of the fluid mesh [185] – in order to mimic the situation of the red blood cell membrane, where the average distance of anchoring points is about 70 nm, much larger than the size of a lipid molecule – and thereby allow, for example, for thermal fluctuations of the distances between neighboring anchoring points. On the other hand, to investigate the behavior of cells on length scales much larger than the mesh size of the spectrin network, it is more efficient to use the same number of bonds for both the fluid and the tethered networks [187]. The simplest case is a harmonic tethering potential, $(1/2)k_{el}(\mathbf{r}_i - \mathbf{r}_j)^2$. This tether network generates a shear modulus $\mu = \sqrt{3}k_{el}$.

11.3 Modeling Membrane Hydrodynamics

Solvent-free models, triangulated surfaces and other discretized curvature models have the disadvantage that they do not contain a solvent, and therefore do not describe the hydrodynamic behavior correctly. However, this apparent disadvantage can be turned into an advantage by combining these models with a mesoscopic hydrodynamics technique. This approach has been employed for dynamically triangulated surfaces [37, 180] and for meshless membrane models in combination with MPC [188], as well as for fixed membrane triangulations in combination with both MPC [187] and the LB method [189].

The solvent particles of the MPC fluid interact with the membrane in two ways to obtain an impermeable membrane with no-slip boundary conditions. First, the membrane vertices are included in the MPC collision procedure, as suggested for polymers in [68], compare Sect. 7.1. Second, the solvent particles are scattered with a bounce-back rule from the membrane surface. Here, solvent particles inside $(1 \leq i \leq N_{in})$ and outside $(N_{in} < i \leq N_s)$ of the vesicle have to be distinguished. The membrane triangles are assumed to have a finite but very small thickness $\delta = 2l_{bs}$. The scattering process is then performed at discrete time steps Δt_{bs}, so that scattering does not occur exactly on the membrane surface, but the solvent particles can penetrate slightly into the membrane film [180]. A similar procedure has been suggested in [26] for spherical colloidal particles embedded in a MPC solvent. Particles which enter the membrane film, i.e., which have a distance to the triangulated surface smaller than l_{bs}, or interior particles which reach the exterior volume and vice versa, are scattered at the membrane triangle with the closest center of mass. Explicitly [180],

$$\mathbf{v}_s^{(new)}(t) = \mathbf{v}_s(t) - \frac{6m_{mb}}{m_s + 3m_{mb}}(\mathbf{v}_s(t) - \mathbf{v}_{tri}(t)) \tag{100}$$

$$\mathbf{v}_{tri}^{(new)}(t) = \mathbf{v}_{tri}(t) + \frac{2m_s}{m_s + 3m_{mb}}(\mathbf{v}_s(t) - \mathbf{v}_{tri}(t)), \tag{101}$$

when $(\mathbf{v}_s(t) - \mathbf{v}_{tri}(t)) \cdot \mathbf{n}_{tri} < 0$, where $\mathbf{v}_s(t)$ and $\mathbf{v}_{tri}(t)$ are the velocities of the solvent particle and of the center of mass of the membrane triangle, respectively, and $\mathbf{n}_{tri}$ is the normal vector of the triangle, which is oriented towards the outside (inside) for external (internal) particles.

The bond flips provide a very convenient way to vary the membrane viscosity η_{mb}, which increases with decreasing probability ψ for the selection of a bond for a bond-flip attempt. The membrane viscosity has been determined quantitatively

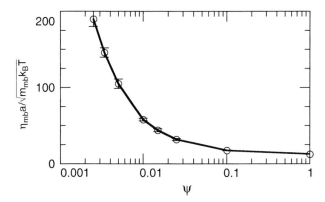

Fig. 24 Dependence of the membrane viscosity η_{mb} on the probability ψ for the selection of a bond for a bond-flip attempt, for a membrane with $N_{mb} = 1,860$ vertices. From [180]

from a simulation of a flat membrane in two-dimensional Poiseuille flow. The triangulated membrane is put in a rectangular box of size $L_x \times L_y$ with periodic boundary conditions in the x-direction. The edge vertices at the lower and upper boundary ($y = \pm L_y/2$) are fixed at their positions. A gravitational force $(m_{mb}g, 0)$ is applied to all membrane vertices to induce a flow. Rescaling of relative velocities is employed as a thermostat. Then, the membrane viscosity is calculated from $\eta_{mb} = \rho_{mb}gL_y/8v_{max}$, where ρ_{mb} is average mass density of the membrane particles, and v_{max} the maximum velocity of the parabolic flow profile. The membrane viscosity η_{mb}, which is obtained in this way, is shown in Fig. 24. As ψ decreases, it takes longer and longer for a membrane particle to escape from the cage of its neighbors, and η_{mb} increases. This is very similar to the behavior of a hard-sphere fluid with increasing density. Finally, for $\psi = 0$, the membrane becomes solid.

11.4 Fluid Vesicles in Shear Flow

The dynamical behavior of *fluid vesicles* in simple shear flow has been studied experimentally [190–193], theoretically [194–201], numerically with the boundary-integral technique [202, 203] or the phase-field method [203, 204], and with meso-scale solvents [37,180,205]. The vesicle shape is now determined by the competition of the curvature elasticity of the membrane, the constraints of constant volume V and constant surface area S, and the external hydrodynamic forces.

Shear flow is characterized (in the absence of vesicles or cells) by the flow field $\mathbf{v} = \dot{\gamma}y\mathbf{e}_x$, where $\mathbf{e}_x$ is a unit vector, compare Sect. 10.4. The control parameter of shear flow is the shear rate $\dot{\gamma}$, which has the dimension of an inverse time. Thus, a dimensionless, scaled shear rate $\dot{\gamma}^* = \dot{\gamma}\tau$ can be defined, where τ is a characteristic relaxation time of a vesicle. Here, $\tau = \eta_0 R_0^3/k_B T$ is used, where η_0 is the solvent viscosity, R_0 the average radius [206]. For $\dot{\gamma}^* < 1$, the internal vesicle dynamics is fast compared to the external perturbation, so that the vesicle shape is hardly affected by the flow field, whereas for $\dot{\gamma}^* > 1$, the flow forces dominate and the vesicle is in a non-equilibrium steady state.

One of the difficulties in theoretical studies of the hydrodynamic effects on vesicle dynamics is the no-slip boundary condition for the embedding fluid on the vesicle surface, which changes its shape dynamically under the effect of flow and curvature forces. In early studies, a fluid vesicle was therefore modeled as an ellipsoid with fixed shape [194]. This simplified model is still very useful as a reference for the interpretation of simulation results.

11.4.1 Generalized Keller–Skalak Theory

The theory of Keller and Skalak [194] describes the hydrodynamic behavior of vesicles of fixed ellipsoidal shape in shear flow, with the viscosities η_{in} and η_0 of the internal and external fluids, respectively. Despite of the approximations needed to derive the equation of motion for the inclination angle θ, which measures the deviation of the symmetry axis of the ellipsoid from the flow direction, this theory describes vesicles in flow surprisingly well. It has been generalized later [197] to describe the effects of a membrane viscosity η_{mb}.

The main result of the theory of Keller and Skalak is the equation of motion for the inclination angle [194],

$$\frac{\mathrm{d}}{\mathrm{d}t}\theta = \frac{1}{2}\dot{\gamma}[-1 + B\cos(2\theta)], \tag{102}$$

where B is a constant, which depends on the geometrical parameters of the ellipsoid, on the viscosity contrast $\eta_{\text{in}}^* = \eta_{\text{in}}/\eta_0$, and the scaled membrane viscosity $\eta_{\text{mb}}^* = \eta_{\text{mb}}/(\eta_0 R_0)$ [180, 194, 197],

$$B = f_0\left[f_1 + \frac{f_1^{-1}}{1 + f_2(\eta_{\text{in}}^* - 1) + f_2 f_3 \eta_{\text{mb}}^*}\right], \tag{103}$$

where f_0, f_1, f_2, and f_3 are geometry-dependent parameters. In the spherical limit, $B \to \infty$. Equation (102) implies the following behavior:

- For $B > 1$, there is a stationary solution, with $\cos(2\theta) = 1/B$. This corresponds to a *tank-treading* motion, in which the orientation of the vesicle axis is time independent, but the membrane itself rotates around the vorticity axis.
- For $B < 1$, no stationary solution exists, and the vesicle shows a *tumbling* motion, very similar to a solid rod-like colloidal particle in shear flow.
- The shear rate $\dot{\gamma}$ only determines the time scale, but does not affect the tank-treading or tumbling behavior. Therefore, a transition between these two types of motion can only be induced by a variation of the vesicle shape or the viscosities.

However, the vesicle shape in shear flow is often not as constant as assumed by Keller and Skalak. In these situations, it is very helpful to compare simulation results with a generalized Keller–Skalak theory, in which shape deformation and thermal fluctuations are taken into account. Therefore, a phenomenological model has been suggested in [180], in which in addition to the inclination angle θ a second parameter is introduced to characterize the vesicle shape and deformation, the asphericity [207]

$$\alpha = \frac{(\lambda_1 - \lambda_2)^2 + (\lambda_2 - \lambda_3)^2 + (\lambda_3 - \lambda_1)^2}{2R_{\mathrm{g}}^4}, \tag{104}$$

where $\lambda_1 \le \lambda_2 \le \lambda_3$ are the eigenvalues of the moment-of-inertia tensor and the squared radius of gyration is $R_{\mathrm{G}}^2 = \lambda_1 + \lambda_2 + \lambda_3$. This implies $\alpha = 0$ for spheres (with $\lambda_1 = \lambda_2 = \lambda_3$), $\alpha = 1$ for long rods (with $\lambda_1 = \lambda_2 \ll \lambda_3$), and $\alpha = 1/4$ for flat disks (with $\lambda_1 \ll \lambda_2 = \lambda_3$). The generalized Keller–Skalak model is then defined by the stochastic equations

$$\zeta_\alpha \frac{\mathrm{d}}{\mathrm{d}t}\alpha = -\partial F/\partial\alpha + A\dot\gamma\sin(2\theta) + \zeta_\alpha g_\alpha(t), \tag{105}$$

$$\frac{\mathrm{d}}{\mathrm{d}t}\theta = \frac{1}{2}\dot\gamma\{-1 + B(\alpha)\cos(2\theta)\} + g_\theta(t), \tag{106}$$

with Gaussian white noises g_α and g_θ, which are determined by

$$\begin{aligned}
\langle g_\alpha(t)\rangle = \langle g_\theta(t)\rangle = \langle g_\alpha(t)g_\theta(t')\rangle &= 0, \\
\langle g_\alpha(t)g_\alpha(t')\rangle &= 2D_\alpha\delta(t-t'), \\
\langle g_\theta(t)g_\theta(t')\rangle &= 2D_\theta\delta(t-t'),
\end{aligned} \tag{107}$$

friction coefficients ζ_α and ζ_θ, and diffusion constants $D_\alpha = k_{\mathrm{B}}T/\zeta_\alpha$ and $D_\theta = k_{\mathrm{B}}T/\zeta_\theta$. Note that ζ_θ does not appear in (106); it drops out because the shear force is also caused by friction.

The form of the stochastic equations (105) and (106) is motivated by the following considerations. The first term in (105), $\partial F/\partial\alpha$, is the thermodynamic force due to bending energy and volume constraints; it is calculated from the free energy $F(\alpha)$. The second term of (105) is the deformation force due to the shear flow. Since the hydrodynamic forces elongate the vesicle for $0 < \theta < \pi/2$ but push to reduce the elongation for $-\pi/2 < \theta < 0$, the flow forces should be proportional to $\sin(2\theta)$ to leading order. The amplitude A is assumed to be independent of the asphericity α. ζ_α and A can be estimated [205] from the results of a perturbation theory [199] in the quasi-spherical limit. Equation (106) is adapted from Keller–Skalak theory. While B is a constant in Keller–Skalak theory, it is now a function of the (time-dependent) asphericity α in (106).

11.4.2 Effects of Membrane Viscosity: Tank-Treading and Tumbling

The theory of Keller and Skalak [194] predicts for fluid vesicles a transition from tank-treading to tumbling with increasing viscosity contrast η_{in}/η_0. This has been confirmed in recent simulations based on a phase-field model [203].

The membrane viscosity η_{mb} is also an important factor for the vesicle dynamics in shear flow. For example, the membrane of red blood cells becomes more viscous on aging [197,208] or in diabetes mellitus [209]. Experiments indicate that the energy dissipation in the membrane is larger than that inside a red blood cell [196,197]. Furthermore, it has been shown recently that vesicles can not only be made from lipid bilayers, but also from bilayers of block copolymers [210]. The membrane viscosity of these "polymersomes" is several orders of magnitude larger than for liposomes, and can be changed over a wide range by varying the polymer chain length [211].

A variation of the membrane viscosity can be implemented easily in dynamically triangulated surface models of membranes, as explained in Sect. 11.2.2. An example of a discocyte in tank-treading motion, which is obtained by such a membrane model [180], is shown in Fig. 25. Simulation results for the inclination angle as a function of the reduced membrane viscosity $\eta_{mb}^* = \eta_{mb}/(\eta_0 R_0)$ are shown in Fig. 26. This demonstrates the tank-treading to tumbling transition of fluid vesicles with increasing membrane viscosity. The threshold shear rate decreases with decreasing reduced volume V^*, since with increasing deviation from the spherical

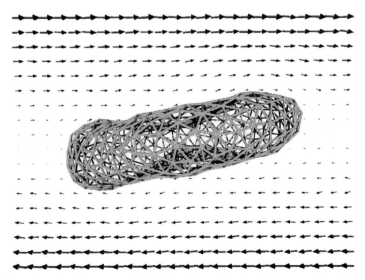

Fig. 25 Snapshot of a discocyte vesicle in shear flow with reduced shear rate $\dot{\gamma}^* = 0.92$, reduced volume $V^* = 0.59$, membrane viscosity $\eta_{mb}^* = 0$, and viscosity contrast $\eta_{in}/\eta_0 = 1$. The *arrows* represent the velocity field in the xz-plane. From [180]

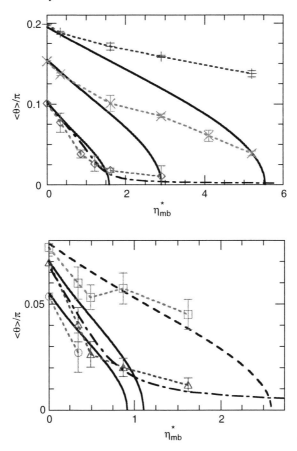

Fig. 26 Average inclination angle $\langle\theta\rangle$ as a function of reduced membrane viscosity η_{mb}^*, for the shear rate $\dot{\gamma}^* = 0.92$ and various reduced volumes V^*. Results are presented for prolate (*circles*) and discocyte (*squares*) vesicles with $V^* = 0.59$, as well as prolate vesicles with $V^* = 0.66$ (*triangles*), 0.78 (*diamonds*), 0.91 (*crosses*), and $V^* = 0.96$ (*pluses*). The *solid and dashed lines* are calculated by K–S theory, (102) and (103), for prolate ($V^* = 0.59$, 0.66, 0.78, 0.91, and 0.96) and oblate ellipsoids ($V^* = 0.59$), respectively. The *dashed-dotted lines* are calculated from (106) with thermal fluctuations, for $V^* = 0.66$, $V^* = 0.78$, and the rotational Peclet number $\dot{\gamma}/D_\theta = 600$ (where D_θ is the rotational diffusion constant). From [180]

shape, the energy dissipation within the membrane increases. Interestingly, the discocyte shape is less affected by the membrane viscosity than the prolate shape for $V^* = 0.59$, since it is more compact – in contrast to a vesicle with viscosity contrast $\eta_{in}/\eta_0 > 1$, where the prolate shape is affected less [180].

Figure 26 also shows a comparison of the simulation data with results of Keller–Skalak (K-S) theory for fixed shape, both without and with thermal fluctuations. Note that there are no adjustable parameters. The agreement of the results of theory and simulations is excellent in the case of vanishing membrane viscosity, $\eta_{mb} = 0$.

For small reduced volumes, $V^* \simeq 0.6$, the tank-treading to tumbling transition is smeared out by thermal fluctuations, with an intermittent tumbling motion occurring in the crossover region. This behavior is captured very well by the generalized K–S model with thermal fluctuations. For larger reduced volumes and non-zero membrane viscosity, significant deviations of theory and simulations become visible. The inclination angle θ is found to decrease much more slowly with increasing membrane viscosity than expected theoretically. This is most likely due to thermal membrane undulations, which are not taken into account in K–S theory.

11.4.3 Swinging of Fluid Vesicles

Recently, a new type of vesicle dynamics in shear flow has been observed experimentally [192], which is characterized by oscillations of the inclination angle θ with $\theta(t) \in [-\theta_0, \theta_0]$ and $\theta_0 < \pi/2$. The vesicles were found to transit from tumbling to this oscillatory motion with increasing shear rate $\dot{\gamma}$. Simultaneously with the experiment, a "vacillating-breathing" mode for quasi-spherical fluid vesicles was predicted theoretically, based on a spherical-harmonics expansion of the equations of motion to leading order (without thermal fluctuations) [199]. This mode exhibits similar dynamical behavior as seen experimentally; however, it "coexists" with the tumbling mode, and its orbit depends on the initial deformation, i.e., it is not a limit cycle. Furthermore, the shear rate appears only as the basic (inverse) time scale, and therefore cannot induce any shape transitions. Hence it does not explain the tumbling-to-oscillatory transition seen in the experiments [192].

Simulation data for the oscillatory mode – which has also been denoted "trembling" [192] or "swinging" [205] mode – are shown in Fig. 27. The simulation results demonstrate that the transition can indeed be induced by increasing shear rate, and that it is robust to thermal fluctuations. Figure 27 also shows that the simulation data are well captured by the generalized K–S model, (105) and (106), which takes into account higher-order contributions in the curvature energy of a vesicle. The theoretical model can therefore be used to predict the full dynamic phase diagram of prolate vesicles as a function of shear rate and membrane viscosity or viscosity contrast, compare Fig. 28. The swinging phase appears at the boundary between the tank-treading and the tumbling phase for sufficiently large shear rates. The phase diagram explains under which conditions the swinging phase can be reached from the tumbling phase with increasing shear rate – as observed experimentally [192].

The generalized K–S model is designed to capture the vesicle flow behavior for non-spherical shapes sufficiently far from a sphere. For quasi-spherical vesicles, a derivation of the equations of motion by a systematic expansion in the undulation amplitudes gives quantitatively more reliable results. An expansion to third order results in a phase diagram [200, 201], which agrees very well with Fig. 28.

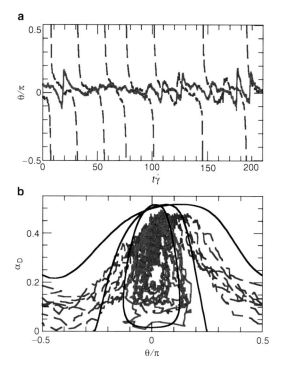

Fig. 27 Temporal evolution of vesicle deformation α_D and inclination angle θ, for $V^* = 0.78$ and $\eta^*_{mb} = 2.9$. Here, $\alpha_D = (L_1 - L_2)/(L_1 + L_2)$, where L_1 and L_2 are the maximum lengths in the direction of the eigenvectors of the gyration tensor in the vorticity plane. The *solid* (*red*) and *dashed* (*blue*) *lines* represent simulation data for $\dot\gamma^* = 3.68$ and 0.92 ($\kappa/k_B T = 10$ and 40, with $\dot\gamma \eta_0 R_0^3/k_B T = 36.8$), respectively. The *solid lines* in (**b**) are obtained from (105), (106) without thermal noise for $\dot\gamma^* = 1.8$, 3.0, and 10 (*from top to bottom*). From [205]

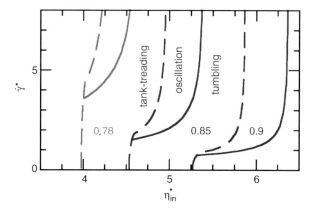

Fig. 28 Dynamical phase diagram as a function of viscosity contrast $\eta^*_{in} = \eta_{in}/\eta_0$, for $\eta^*_{mb} = 0$ and various reduced volumes V^*, calculated from (105), (106) without thermal noise. The tank-treading phase is located on the left-hand side of the *dashed lines*. The *solid lines* represent the tumbling-to-swinging transitions. From [205]

11.4.4 Flow-Induced Shape Transformations

Shear flow does not only induce different dynamical modes of prolate and oblate fluid vesicles, it can also induce phase transformations. The simplest case is a oblate fluid vesicle with $\eta_{mb} = 0$ and viscosity contrast $\eta_{in}/\eta_0 = 1$. When the reduced shear rate reaches $\dot{\gamma}^* \simeq 1$, the discocyte vesicles are stretched by the flow forces into a prolate shape [37,180,202]. A similar transition is found for stomatocyte vesicles, except that in this case a larger shear rate $\dot{\gamma}^* \simeq 3$ is required. In the case of non-zero membrane viscosity, a rich phase behavior appears, see Fig. 29.

Surprisingly, flow forces can not only stretch vesicles into a more elongated shape, but can also induce a transition from an elongated prolate shape into a more compact discocyte shape [180]. Simulation results for the latter transition are shown in Fig. 30. This transition is possible, because in a range of membrane viscosities, the prolate shape is in the tumbling phase, while the oblate shape is tank-treading, compare Fig. 26. Of course, this requires that the free energies of the two shapes are nearly equal, which implies a reduced volume of $V^* \simeq 0.6$. Thus, a prolate vesicle in this regime starts tumbling; as the inclination angle becomes negative, shear forces push to shrink the long axis of the vesicle; when this force is strong enough to overcome the free-energy barrier between the prolate and the oblate phase, a shape transformation can be induced, compare Fig. 30. The vesicle then remains in the stable tank-treading state.

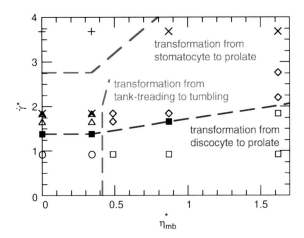

Fig. 29 Dynamical phase diagram of a vesicle in shear flow for reduced volume $V^* = 0.59$. Symbols correspond to simulated parameter values, and indicate tank-treading discocyte and tank-treading prolate (*circles*), tank-treading prolate and unstable discocyte (*triangles*), tank-treading discocyte and tumbling (transient) prolate (*open squares*), tumbling with shape oscillation (*diamonds*), unstable stomatocyte (*pluses*), stable stomatocyte (*crosses*), and near transition (*filled squares*). The *dashed lines* are guides to the eye. From [180]

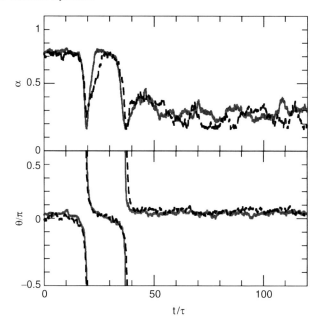

Fig. 30 Time dependence of asphericity α and inclination angle θ, for $\dot{\gamma}^* = 1.84$, $\eta^*_{mb} = 1.62$, and $V^* = 0.59$. The *dashed lines* are obtained from (105) and (106), with $\zeta_\alpha = 100$, $A = 12$, and $B(\alpha) = 1.1 - 0.17\alpha$. From [212]

At higher shear rates, an intermittent behavior has been observed, in which the vesicle motion changes in irregular intervals between tumbling and tank-treading [180].

11.4.5   Vesicle Fluctuations Under Flow in Two Dimensions

At finite temperature, stochastic fluctuations of the membrane due to thermal motion affect the dynamics of vesicles. Since the calculation of thermal fluctuations under flow conditions requires long times and large membrane sizes (in order to have a sufficient range of undulation wave vectors), simulations have been performed for a two-dimensional system in the stationary tank-treading state [213]. For comparison, in the limit of small deviations from a circle, Langevin-type equations of motion have been derived, which are highly nonlinear due to the constraint of constant perimeter length [213].

The effect of the shear flow is to induce a tension in the membrane, which reduces the amplitude of thermal membrane undulations. This tension can be extracted directly from simulation data for the undulation spectrum. The reduction of the undulation amplitudes also implies that the fluctuations of the inclination angle θ get reduced with increasing shear rate. The theory for quasi-circular shapes predicts a universal behavior as a function of the scaled shear rate $\dot{\gamma}^* \Delta^{1/2} \kappa / (R_0 k_B T)$, where

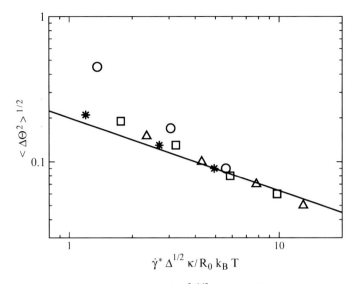

Fig. 31 Fluctuations of the inclination angle $\langle \Delta\theta^2\rangle^{1/2}$ of a two-dimensional fluctuating vesicle in shear flow, as a function of scaled shear rate $\dot{\gamma}^*\Delta^{1/2}\kappa/(R_0 k_B T)$, where Δ is the dimensionless excess membrane length. Symbols indicate simulation data for different internal vesicle areas A for fixed membrane length, with $A^* \equiv A/\pi R_0^2 = 0.95$ (*squares*), $A^* = 0.90$ (*triangles*), $A^* = 0.85$ (*stars*), and $A^* = 0.7$ (*circles*). The *solid line* is the theoretical result in the quasi-circular limit. From [213]

$\dot{\gamma}^* = \dot{\gamma}\eta_0 R_0^3/\kappa$ is the reduced shear rate in two dimensions, and Δ is the dimension-less excess membrane length. Theory and simulation results for the inclination angle as a function of the reduced shear rate are shown in Fig. 31. There are no adjustable parameters. The agreement is excellent as long as the deviations from the circular shape are not too large [213].

11.5 Fluid Vesicles and Red Blood Cells in Capillary Flow

11.5.1 RBC Deformation in Narrow Capillaries

The deformation of single RBCs and single fluid vesicles in capillary flows were studied theoretically by lubrication theories [214–216] and boundary-integral methods [217–219]. In most of these studies, axisymmetric shapes which are coaxial with the center of the capillary were assumed and cylindrical coordinates were employed. In order to investigate non-axisymmetric shapes as well as flow-induced shape transformations, a fully three-dimensional simulation approach is required.

We focus here on the behavior of single red blood cells in capillary flow [187], as described by a triangulated surface model for the membrane (compare Sect. 11.2.4) immersed in a MPC solvent (see Sect. 11.3). The radius of the capillary, R_{cap}, is taken to be slightly larger than the mean vesicle or RBC radius, $R_0 = \sqrt{S/4\pi}$, where S is the membrane area. Snapshots of vesicle and RBC shapes in flow are shown in Fig. 32 for a reduced volume of $V^* = 0.59$, where the vesicle shape at rest is a discocyte. For sufficiently small flow velocities, the discocyte shape is retained. However, the discocyte is found *not* in a coaxial orientation; instead the shortest eigenvalue of the gyration tensor is oriented perpendicular to the cylinder axis [187]. Since two opposite sides of the rim of the discocyte are closer to the wall where the flow velocity is small, the rotational symmetry is slightly disturbed and the top view looks somewhat triangular, see Fig. 32a. With increasing flow velocity, a shape transition to an axisymmetric shape occurs. In the case of fluid vesicles this is a

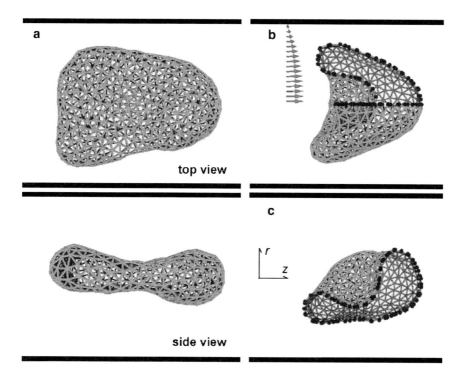

Fig. 32 Snapshots of vesicles in capillary flow, with bending rigidity $\kappa/k_B T = 20$ and capillary radius $R_{cap} = 1.4R_0$. **a** Fluid vesicle with discoidal shape at the mean fluid velocity $v_m \tau/R_{cap} = 41$, both in side and top views. **b** Elastic vesicle (RBC model) with parachute shape at $v_m \tau/R_{cap} = 218$ (with shear modulus $\mu R_0^2/k_B T = 110$). The *blue arrows* represent the velocity field of the solvent. **c** Elastic vesicle with slipper-like shape at $v_m \tau/R_{cap} = 80$ (with $\mu R_0^2/k_B T = 110$). The inside and outside of the membrane are depicted in *red* and *green*, respectively. The upper front quarter of the vesicle in (**b**) and the front half of the vesicle in (**c**) are removed to allow for a look into the interior; the *black circles* indicate the lines where the membrane has been cut in this procedure. *Thick black lines* indicate the walls of the cylindrical capillary. From [187]

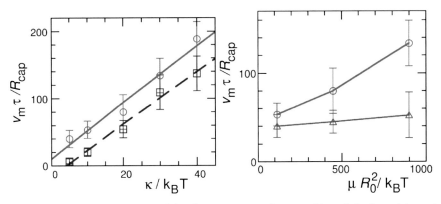

Fig. 33 Critical flow velocity v_m of the discocyte-to-parachute transition of elastic vesicles and of the discocyte-to-prolate transition of fluid vesicles, as a function of the bending rigidity for $\mu R_0^2/k_B T = 110$ (left), and of the shear modulus μ for $\kappa/k_B T = 10$ (right). From [187]

prolate shape, while in the case of RBCs a parachute shape is found. Such parachute shapes of red blood cells have previously been observed experimentally [209, 220].

The fundamental difference between the flow behaviors of fluid vesicles and red blood cells at high flow velocities is due to the shear elasticity of the spectrin network. Its main effect for $\mu R_0^2/\kappa \gtrsim 1$ is to suppress the discocyte-to-prolate transition, because the prolate shape would acquire an elastic energy of order μR_0^2. In comparison, the shear stress in the parachute shape is much smaller.

Some diseases, such as diabetes mellitus and sickle cell anemia, change the mechanical properties of RBCs; a reduction of RBC deformability was found to be responsible for an enhanced flow resistance of blood [221]. Therefore, it is very important to understand the relation of RBC elasticity and flow properties in capillaries. The flow velocity at the discocyte-to-prolate transition of fluid vesicles and at the discocyte-to-parachute transition is shown in Fig. 33 as a function of the bending rigidity and the shear modulus, respectively. In both cases, an approximately linear dependence is obtained [187],

$$v_m^c \frac{\tau}{R_{\text{cap}}} = 0.1 \frac{\mu R_0^2}{k_B T} + 4.0 \frac{\kappa}{k_B T}. \tag{108}$$

This result suggests that parachute shapes of RBCs should appear for flow velocities larger than $v_m^c = 800 R_{\text{cap}}/\tau \simeq 0.2 \, \text{mm s}^{-1}$ under physiological conditions. This is consistent with the experimental results of [222], and is in the range of microcirculation in the human body.

Figure 33 (right) also shows that there is a metastable region, where discocytes are seen for increasing flow velocity, but parachute shapes for decreasing flow velocity. This hysteresis becomes more pronounced with increasing shear modulus. It is believed to be due to a suppression of thermal fluctuations with increasing $\mu R_0^2/k_B T$.

11.5.2 Flow in Wider Capillaries

The flow of many red blood cells in wider capillaries has also been investigated by several simulation techniques. Discrete fluid-particle simulations – an extension of DPD – in combination with bulk-elastic discocyte cells (in contrast to the membrane elasticity of real red blood cells) have been employed to investigate the dynamical clustering of red blood cells in capillary vessels [223, 224]. An immersed finite-element model – a combination of the immersed boundary method for the solvent hydrodynamics [225] and a finite-element method to describe the membrane elasticity – has been developed to study red blood cell aggregation [226]. Finally, it has been demonstrated that the LB method for the solvent in combination with a triangulated mesh model with curvature and shear elasticity for the membrane can be used efficiently to simulate RBC suspensions in wider capillaries [189].

12 Viscoelastic Fluids

One of the unique properties of soft matter is its viscoelastic behavior [13]. Because of the long structural relaxation times, the internal degrees of freedom cannot relax sufficiently fast in an oscillatory shear flow already at moderate frequencies, so that there is an elastic restoring force which pushes the system back towards its previous state. Well-studied examples of viscoelastic fluids are polymer solutions and polymer melts [6, 13].

The viscoelastic behavior of polymer solutions leads to many unusual flow phenomena, such as viscoelastic phase separation [227]. There is also a second level of complexity in soft matter systems, in which a colloidal component is dispersed in a solvent, which is itself a complex fluid. Examples are spherical or rod-like colloids dispersed in polymer solutions. Shear flow can induce particle aggregation and alignment in these systems [228].

It is therefore desirable to generalize the MPC technique to model viscoelastic fluids, while retaining as much as possible of the computational efficiency of standard MPC for Newtonian fluids. This can be done by replacing the point particles of standard MPC by harmonic dumbbells with spring constant K [229].

As for point particles, the MPC algorithm consists of two steps, streaming and collisions. In the streaming step, within a time interval Δt, the motion of all dumbbells is governed by Newton's equations of motion. The center-of-mass coordinate of each dumbbell follows a simple ballistic trajectory. The evolution of the relative coordinates of dumbbell i, which consists of two monomers at positions $\mathbf{r}_{i1}(t)$ and $\mathbf{r}_{i2}(t)$ with velocities $\mathbf{v}_{i1}(t)$ and $\mathbf{v}_{i2}(t)$, respectively, is determined by the harmonic interaction potential, so that

$$\mathbf{r}_{i1}(t+\Delta t) - \mathbf{r}_{i2}(t+\Delta t) = \mathbf{A}_i(t)\cos(\omega_0\Delta t) + \mathbf{B}_i(t)\sin(\omega_0\Delta t); \quad (109)$$
$$\mathbf{v}_{i1}(t+\Delta t) - \mathbf{v}_{i2}(t+\Delta t) = -\omega_0\mathbf{A}_i(t)\sin(\omega_0\Delta t)$$

$$+\omega_0 \mathbf{B}_i(t)\cos(\omega_0 \Delta t), \tag{110}$$

with angular frequency $\omega_0 = \sqrt{2K/m}$. The amplitudes $\mathbf{A}_i(t)$ and $\mathbf{B}_i(t)$ are determined by the initial positions and velocities at time t. The collision step is performed for the two point particles constituting a dumbbell in exactly the same way as for MPC point-particle fluids. This implies, in particular, that the various collision rules of MPC, such as SRD, AT$-a$ or AT$+a$, can all be employed also for simulations of viscoelastic solvents, depending on the requirements of the system under consideration. Since the streaming step is only a little more time consuming and the collision step is identical, simulations of the viscoelastic MPC fluid can be performed with essentially the same efficiency as for the standard point-particle fluid.

The behavior of harmonic dumbbells in dilute solution has been studied in detail analytically [230]. These results can be used to predict the zero-shear viscosity η and the storage and loss moduli, $G'(\omega)$ and $G''(\omega)$ in oscillatory shear with frequency ω, of the MPC dumbbell fluid. This requires the solvent viscosity and diffusion constant of monomers in the solvent. Since the viscoelastic MPC fluid consists of dumbbells only, the natural assumption is to employ the viscosity η_{MPC} and diffusion constant D of an MPC point-particle fluid of the same density. The zero-shear viscosity is then found to be [229]

$$\eta = \eta_{MPC} + \frac{\rho}{2}\frac{k_B T}{\omega_H}, \tag{111}$$

where

$$\omega_H = \frac{4K}{\zeta} = \frac{4DK}{k_B T}. \tag{112}$$

Similarly, the storage and loss modulus, and the average dumbbell extension, are predicted to be [229]

$$G' = \frac{\rho k_B T}{2}\frac{(\omega/\omega_H)^2}{1+(\omega/\omega_H)^2}, \tag{113}$$

$$G'' = \eta_{MPC}\,\omega + \frac{\rho k_B T}{2}\frac{\omega/\omega_H}{1+(\omega/\omega_H)^2}, \tag{114}$$

and

$$\frac{\langle r^2 \rangle}{\langle r^2 \rangle_{eq}} = 1 + \frac{2}{3}(\dot{\gamma}/\omega_H)^2. \tag{115}$$

Simulation data are shown in Fig. 34, together with the theoretical predications (113) and (114). The comparison shows a very good agreement. This includes not only the linear and quadratic frequency dependence of G'' and G' for small ω, respectively, but also the leveling off when ω reaches ω_H. In case of G'', there is quantitative agreement without any adjustable parameters, whereas G' is somewhat overestimated by (113) for small spring constants K. The good agreement of theory and simulations implies that the characteristic frequency decreases linearly with de-

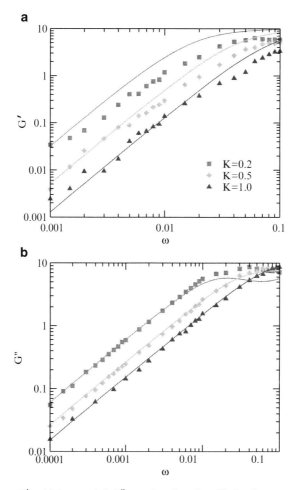

Fig. 34 a Storage G' and **b** loss moduli G'', as a function of oscillation frequency ω on a double-logarithmic scale, for systems of dumbbells with various spring constants ranging from $K = 0.2$ to $K = 1.0$. Simulations are performed in two dimensions with the SRD collision rule. The wall separation and the collision time are $L_y = 10$ and $\Delta t = 0.02$, respectively. From [229]

creasing spring constant K and mean free path λ (since $D \sim \lambda$). A comparison of other quantities, such as the zero-shear viscosity, shows a similar quantitative agreement [229].

13 Conclusions and Outlook

In the short time since Malevanets and Kapral introduced MPC dynamics as a particle-based mesoscale simulation technique for studying the hydrodynamics of complex fluids, there has been enormous progress. It has been shown that kinetic theory can be generalized to calculate transport coefficients, several collision algorithms have been proposed and employed, and the method has been generalized to describe multi-phase flows and viscoelastic fluids. The primary applications to date – which include studies of the equilibrium dynamics and flow properties of colloids, polymers, and vesicles in solution – have dealt with mesoscopic particles embedded in a single-component Newtonian solvent. An important advantage of this algorithm is that it is very straightforward to model the dynamics for the embedded particles using a hybrid MPC–MD simulations approach. The results of these studies are in excellent quantitative agreement with both theoretical predictions and results obtained using other simulation techniques.

How will the method develop in the future? This is of course difficult to predict. However, it seems clear that there will be two main directions, a further development of the method itself, and its application to new problems in Soft Matter hydrodynamics. On the methodological front, there are several very recent developments, like angular-momentum conservation, multi-phase flows and viscoelastic fluids, which have to be explored in more detail. It will also be interesting to combine them to study, for example, multi-phase flows of viscoelastic fluids. On the application side, the trend will undoubtedly be towards more complex systems, in which thermal fluctuations are important. In such systems, the method can play out its strengths, because the interactions of colloids, polymers, and membranes with the mesoscale solvent can all be treated on the same basis.

Acknowledgments Financial support from the Donors of the American Chemical Society Petroleum Research Fund, the National Science Foundation under grant No. DMR-0513393, DMR-0706017, the German Research Foundation (DFG) within the SFB TR6 "Physics of Colloidal Dispersion in External Fields," and the Priority Program "Nano- and Microfluidics" are gratefully acknowledged. We thank Elshad Allahyarov, Luigi Cannavacciuolo, Jens Elgeti, Reimar Finken, Ingo Götze, Jens Harting, Martin Hecht, Hans Herrmann, Antonio Lamura, Kiaresch Mussawisade, Hiroshi Noguchi, Guoai Pan, Marisol Ripoll, Udo Seifert, Yu-Guo Tao, and Erkan Tüzel for many stimulating discussions and enjoyable collaborations.

References

1. P. J. Hoogerbrugge and J. M. V. A. Koelman, Europhys. Lett. **19**, 155 (1992).
2. P. Espanol, Phys. Rev. E **52**, 1734 (1995).
3. P. Espanol and P. B. Warren, Europhys. Lett. **30**, 191 (1995).
4. G. R. McNamara and G. Zanetti, Phys. Rev. Lett. **61**, 2332 (1988).
5. X. Shan and H. Chen, Phys. Rev. E **47**, 1815 (1993).
6. X. He and L.-S. Luo, Phys. Rev. E **56**, 6811 (1997).
7. G. A. Bird, *Molecular Gas Dynamics and the Direct Simulation of Gas Flows* (Oxford University Press, Oxford, 1994).
8. F. J. Alexander and A. L. Garcia, Comp. in Phys. **11**, 588 (1997).
9. A. L. Garcia, *Numerical Methods for Physics* (Prentice Hall, 2000).
10. U. Frisch, B. Hasslacher, and Y. Pomeau, Phys. Rev. Lett. **56**, 1505 (1986).
11. R. Adhikari, K. Stratford, M. E. Cates, and A. J. Wagner, Europhys. Lett. **71**, 473 (2005).

12. J. K. G. Dhont, *An Introduction to Dynamics of Colloids* (Elsevier, Amsterdam, 1996).
13. R. G. Larson, *The Structure and Rheology of Complex Fluids* (Oxford University Press, Oxford, 1999).
14. M. Ripoll, K. Mussawisade, R. G. Winkler, and G. Gompper, Europhys. Lett. **68**, 106 (2004).
15. M. Ripoll, K. Mussawisade, R. G. Winkler, and G. Gompper, Phys. Rev. E **72**, 016701 (2005).
16. M. Hecht, J. Harting, T. Ihle, and H. J. Herrmann, Phys. Rev. E **72**, 011408 (2005).
17. J. T. Padding and A. A. Louis, Phys. Rev. E **74**, 031402 (2006).
18. A. Malevanets and R. Kapral, J. Chem. Phys. **110**, 8605 (1999).
19. A. Malevanets and R. Kapral, J. Chem. Phys. **112**, 7260 (2000).
20. T. Ihle and D. M. Kroll, Phys. Rev. E **67**, 066705 (2003).
21. T. Ihle and D. M. Kroll, Phys. Rev. E **63**, 020201(R) (2001).
22. A. Mohan and P. S. Doyle, Macromolecules **40**, 4301 (2007).
23. J. M. Kim and P. S. Doyle, J. Chem. Phys. **125**, 074906 (2006).
24. N. Kikuchi, A. Gent, and J. M. Yeomans, Eur. Phys. J. E **9**, 63 (2002).
25. M. Ripoll, R. G. Winkler, and G. Gompper, Eur. Phys. J. E **23**, 349 (2007).
26. N. Kikuchi, C. M. Pooley, J. F. Ryder, and J. M. Yeomans, J. Chem. Phys. **119**, 6388 (2003).
27. T. Ihle and D. M. Kroll, Phys. Rev. E **67**, 066706 (2003).
28. T. Ihle, E. Tüzel, and D. M. Kroll, Phys. Rev. E **72**, 046707 (2005).
29. J. A. Backer, C. P. Lowe, H. C. J. Hoefsloot, and P. D. Iedema, J. Chem. Phys. **122**, 1 (2005).
30. R. Kapral, Adv. Chem. Phys., to appear (2008).
31. E. Allahyarov and G. Gompper, Phys. Rev. E **66**, 036702 (2002).
32. N. Noguchi, N. Kikuchi, and G. Gompper, Europhys. Lett. **78**, 10005 (2007).
33. T. Ihle, E. Tüzel, and D. M. Kroll, Europhys. Lett. **73**, 664 (2006).
34. E. Tüzel, T. Ihle, and D. M. Kroll, Math. Comput. Simulat. **72**, 232 (2006).
35. C. M. Pooley and J. M. Yeomans, J. Phys. Chem. B **109**, 6505 (2005).
36. A. Lamura, G. Gompper, T. Ihle, and D. M. Kroll, Europhys. Lett. **56**, 319 (2001).
37. H. Noguchi and G. Gompper, Phys. Rev. Lett. **93**, 258102 (2004).
38. I. O. Götze, H. Noguchi, and G. Gompper, Phys. Rev. E **76**, 046705 (2007).
39. J. F. Ryder, *Mesoscopic Simulations of Complex Fluids*, Ph.D. thesis, University of Oxford (2005).
40. I. O. Götze, private communication (2007).
41. J. Erpenbeck, Phys. Rev. Lett. **52**, 1333 (1984).
42. M. P. Allen and D. J. Tildesley, *Computer Simulation of Liquids* (Clarendon Press, Oxford, 1987).
43. J. H. Irving and J. G. Kirkwood, J. Chem. Phys. **18**, 817 (1950).
44. H. T. Davis, *Statistical Mechanics of Phases, Interfaces, and Thin Films* (VCH Publishers, Inc., 1996).
45. E. Tüzel, G. Pan, T. Ihle, and D. M. Kroll, Europhys. Lett. **80**, 40010 (2007).
46. F. Reif, *Fundamentals of Statistical and Thermal Physics* (Mc-Graw Hill, 1965).
47. R. Zwanzig, *Lectures in Theoretical Physics*, vol. 3 (Wiley, New York, 1961).
48. H. Mori, Prog. Theor. Phys. **33**, 423 (1965).
49. H. Mori, Prog. Theor. Phys. **34**, 399 (1965).
50. J. W. Dufty and M. H. Ernst, J. Phys. Chem. **93**, 7015 (1989).
51. T. Ihle, E. Tüzel, and D. M. Kroll, Phys. Rev. E **70**, 035701(R) (2004).
52. C. M. Pooley, *Mesoscopic Modelling Techniques for Complex Fluids*, Ph.D. thesis, University of Oxford (2003).
53. E. Tüzel, M. Strauss, T. Ihle, and D. M. Kroll, Phys. Rev. E **68**, 036701 (2003).
54. E. Tüzel, *Particle-based mesoscale modeling of flow and transport in complex fluids*, Ph.D. thesis, University of Minnesota (2006).
55. T. Ihle and E. Tüzel, Prog. Comp. Fluid Dynamics **8**, 123 (2008).
56. B. J. Berne and R. Pecora, *Dynamic Light Scattering: With Applications to Chemistry, Biology and Physics* (Wiley, New York, 1976).
57. E. Tüzel, T. Ihle, and D. M. Kroll, Phys. Rev. E **74**, 056702 (2006).
58. H. Noguchi and G. Gompper, Europhys. Lett. **79**, 36002 (2007).
59. H. Noguchi and G. Gompper, Phys. Rev. E **78**, 016706 (2008).

60. T. Ihle, J. Phys.: Condens. Matter **20**, 235224 (2008).
61. H. B. Callen, *Thermodynamics* (Wiley, New York, 1960).
62. Y. Hashimoto, Y. Chen, and H. Ohashi, Comp. Phys. Commun. **129**, 56 (2000).
63. Y. Inoue, Y. Chen, and H. Ohashi, Comp. Phys. Commun. **201**, 191 (2004).
64. Y. Inoue, S. Takagi, and Y. Matsumoto, Comp. Fluids **35**, 971 (2006).
65. T. Sakai, Y. Chen, and H. Ohashi, Comp. Phys. Commun. **129**, 75 (2000).
66. T. Sakai, Y. Chen, and H. Ohashi, Phys. Rev. E **65**, 031503 (2002).
67. T. Sakai, Y. Chen, and H. Ohashi, Colloids Surf., A **201**, 297 (2002).
68. A. Malevanets and J. M. Yeomans, Europhys. Lett. **52**, 231 (2000).
69. J. Elgeti and G. Gompper, in *NIC Symposium 2008*, edited by G. Münster, D. Wolf, and M. Kremer (Neumann Institute for Computing, Jülich, 2008), vol. 39 of *NIC series*, pp. 53–61.
70. M. Doi and S. F. Edwards, *The Theory of Polymer Dynamics* (Oxford University Press, Oxford, 1986).
71. E. Falck, O. Punkkinen, I. Vattulainen, and T. Ala-Nissila, Phys. Rev. E **68**, 050102(R) (2003).
72. R. G. Winkler, K. Mussawisade, M. Ripoll, and G. Gompper, J. Phys.: Condens. Matter **16**, S3941 (2004).
73. K. Mussawisade, M. Ripoll, R. G. Winkler, and G. Gompper, J. Chem. Phys. **123**, 144905 (2005).
74. M. A. Webster and J. M. Yeomans, J. Chem. Phys. **122**, 164903 (2005).
75. E. Falck, J. M. Lahtinen, I. Vattulainen, and T. Ala-Nissila, Eur. Phys. J. E **13**, 267 (2004).
76. M. Hecht, J. Harting, and H. J. Herrmann, Phys. Rev. E **75**, 051404 (2007).
77. S. H. Lee and R. Kapral, J. Chem. Phys. **121**, 11163 (2004).
78. Y. Inoue, Y. Chen, and H. Ohashi, J. Stat. Phys. **107**, 85 (2002).
79. J. T. Padding, A. Wysocki, H. Löwen, and A. A. Louis, J. Phys.: Condens. Matter **17**, S3393 (2005).
80. S. H. Lee and R. Kapral, Physica A **298**, 56 (2001).
81. A. Lamura and G. Gompper, Eur. Phys. J. E **9**, 477 (2002).
82. J. T. Padding, private communication (2007).
83. C. Pierleoni and J.-P. Ryckaert, Phys. Rev. Lett. **61**, 2992 (1991).
84. B. Dünweg and K. Kremer, Phys. Rev. Lett. **66**, 2996 (1991).
85. C. Pierleoni and J.-P. Ryckaert, J. Chem. Phys. **96**, 8539 (1992).
86. B. Dünweg and K. Kremer, J. Chem. Phys. **99**, 6983 (1993).
87. C. Aust, M. Kröger, and S. Hess, Macromolecules **32**, 5660 (1999).
88. E. S. Boek, P. V. Coveney, H. N. W. Lekkerkerker, and P. van der Schoot, Phys. Rev. E **55**, 3124 (1997).
89. P. Ahlrichs and B. Dünweg, Int. J. Mod. Phys. C **9**, 1429 (1998).
90. P. Ahlrichs, R. Everaers, and B. Dünweg, Phys. Rev. E **64**, 040501 (2001).
91. N. A. Spenley, Europhys. Lett. **49**, 534 (2000).
92. C. P. Lowe, A. F. Bakker, and M. W. Dreischor, Europhys. Lett. **67**, 397 (2004).
93. G. K. Batchelor, J. Fluid Mech. **52**, 245 (1972).
94. A. J. C. Ladd, Phys. Fluids **9**, 481 (1997).
95. K. Höfler and S. Schwarzer, Phys. Rev. E **61**, 7146 (2000).
96. J. T. Padding and A. A. Louis, Phys. Rev. Lett. **93**, 220601 (2004).
97. H. Hayakawa and K. Ichiki, Phys. Rev. E **51**, R3815 (1995).
98. H. Hayakawa and K. Ichiki, Phys. Fluids **9**, 481 (1997).
99. M. H. Ernst, E. H. Hauge, and J. M. J. van Leeuwen, Phys. Rev. Lett. **25**, 1254 (1970).
100. M. Hecht, J. Harting, M. Bier, J. Reinshagen, and H. J. Herrmann, Phys. Rev. E **74**, 021403 (2006).
101. B. V. Derjaguin and D. P. Landau, Acta Phys. Chim. **14**, 633 (1941).
102. W. B. Russel, D. A. Saville, and W. Schowalter, *Colloidal dispersions* (Cambridge University Press, Cambridge, 1995).
103. S. H. Lee and R. Kapral, J. Chem. Phys. **122**, 214916 (2005).
104. J. G. Kirkwood and J. Riseman, J. Chem. Phys. **16**, 565 (1948).
105. J. P. Erpenbeck and J. G. Kirkwood, J. Chem. Phys. **29**, 909 (1958).

106. E. P. Petrov, T. Ohrt, R. G. Winkler, and P. Schwille, Phys. Rev. Lett. **97**, 258101 (2006).
107. B. H. Zimm, J. Chem. Phys. **24**, 269 (1956).
108. S. H. Lee and R. Kapral, J. Chem. Phys. **124**, 214901 (2006).
109. M. H. Ernst, E. H. Hauge, and J. M. J. van Leeuwen, Phys. Rev. A **4**, 2055 (1971).
110. J.-P. Hansen and I. R. McDonald, *Theory of Simple Liquids* (Academic Press, London, 1986).
111. B. Liu and B. Dünweg, J. Chem. Phys. **118**, 8061 (2003).
112. P. Ahlrichs and B. Dünweg, J. Chem. Phys. **111**, 8225 (1999).
113. M. Fixmann, J. Chem. Phys. **78**, 1594 (1983).
114. M. Schmidt and W. Burchard, Macromolecules **14**, 210 (1981).
115. W. H. Stockmayer and B. Hammouda, Pure & Appl. Chem. **56**, 1373 (1984).
116. P. E. Rouse, J. Chem. Phys. **21**, 1272 (1953).
117. P. G. de Gennes, *Scaling Concepts in Polymer Physics* (Cornell University, Ithaca, 1979).
118. J. des Cloizeaux and G. Jannink, *Polymer Solutions: Their Modelling and Structure* (Clarendon Press, Oxford, 1990).
119. D. Ceperly, M. H. Kalos, and J. L. Lebowitz, Macromolecules **14**, 1472 (1981).
120. K. Kremer and K. Binder, Comput. Phys. Rep. **7**, 261 (1988).
121. R. G. Winkler, L. Harnau, and P. Reineker, Macromol. Theory Simul. **6**, 1007 (1997).
122. R. Chang and A. Yethiraj, J. Chem. Phys. **114**, 7688 (2001).
123. N. Kikuchi, A. Gent, and J. M. Yeomans, Eur. Phys. J. E **9**, 63 (2002).
124. N. Kikuchi, J. F. Ryder, C. M. Pooley, and J. M. Yeomans, Phys. Rev. E **71**, 061804 (2005).
125. I. Ali and J. M. Yeomans, J. Chem. Phys. **123**, 234903 (2005).
126. I. Ali and J. M. Yeomans, J. Chem. Phys. **121**, 8635 (2004).
127. I. Ali, D. Marenduzzo, and J. F. D. Yeomans, Phys. Rev. Lett. **96**, 208102 (2006).
128. N. Watari, M. Makino, N. Kikuchi, R. G. Larson, and M. Doi, J. Chem. Phys. **126**, 094902 (2007).
129. F. Brochard-Wyart, Europhys. Lett. **23**, 105 (1993).
130. F. Brochard-Wyart, H. Hervet, and P. Pincus, Europhys. Lett. **26**, 511 (1994).
131. F. Brochard-Wyart, Europhys. Lett. **30**, 210 (1995).
132. L. Cannavacciuolo, R. G. Winkler, and G. Gompper, EPL **83**, 34007 (2008).
133. U. S. Agarwal, A. Dutta, and R. A. Mashelkar, Chem. Eng. Sci. **49**, 1693 (1994).
134. R. M. Jendrejack, D. C. Schwartz, J. J. de Pablo, and M. D. Graham, J. Chem. Phys. **120**, 2513 (2004).
135. O. B. Usta, J. E. Butler, and A. J. C. Ladd, Phys. Fluids **18**, 031703 (2006).
136. R. Khare, M. D. Graham, and J. J. de Pablo, Phys. Rev. Lett. **96**, 224505 (2006).
137. D. Stein, F. H. J. van der Heyden, W. J. A. Koopmans, and C. Dekker, Proc. Natl. Acad. Sci. USA **103**, 15853 (2006).
138. O. B. Usta, J. E. Butler, and A. J. C. Ladd, Phys. Rev. Lett. **98**, 098301 (2007).
139. G. S. Grest, K. Kremer, and T. A. Witten, Macromolecules **20**, 1376 (1987).
140. C. N. Likos, Phys. Rep. **348**, 267 (2001).
141. D. Vlassopoulos, G. Fytas, T. Pakula, and J. Roovers, J. Phys.: Condens. Matter **13**, R855 (2001).
142. M. Ripoll, R. G. Winkler, and G. Gompper, Phys. Rev. Lett. **96**, 188302 (2006).
143. G. S. Grest, K. Kremer, S. T. Milner, and T. A. Witten, Macromolecules **22**, 1904 (1989).
144. D. R. Mikulencak and J. F. Morris, J. Fluid Mech. **520**, 215 (2004).
145. A. Link and J. Springer, Macromolecules **26**, 464 (1993).
146. R. E. Teixeira, H. P. Babcock, E. S. G. Shaqfeh, and S. Chu, Macromolecules **38**, 581 (2005).
147. R. G. Winkler, Phys. Rev. Lett. **97**, 128301 (2006).
148. P. G. de Gennes, J. Chem. Phys. **60**, 5030 (1974).
149. P. LeDuc, C. Haber, G. Bao, and D. Wirtz, Nature (London) **399**, 564 (1999).
150. D. E. Smith, H. P. Babcock, and S. Chu, Science **283**, 1724 (1999).
151. C. Aust, M. Kröger, and S. Hess, Macromolecules **35**, 8621 (2002).
152. Y. Navot, Phys. Fluids **10**, 1819 (1998).
153. R. Goetz and R. Lipowsky, J. Chem. Phys. **108**, 7397 (1998).
154. R. Goetz, G. Gompper, and R. Lipowsky, Phys. Rev. Lett. **82**, 221 (1999).
155. W. K. den Otter and W. J. Briels, J. Chem. Phys. **118**, 4712 (2003).

156. J. C. Shillcock and R. Lipowsky, J. Chem. Phys. **117**, 5048 (2002).
157. L. Rekvig, B. Hafskjold, and B. Smit, Phys. Rev. Lett. **92**, 116101 (2004).
158. M. Laradji and P. B. S. Kumar, Phys. Rev. Lett. **93**, 198105 (2004).
159. V. Ortiz, S. O. Nielsen, D. E. Discher, M. L. Klein, R. Lipowsky, and J. Shillcock, J. Phys. Chem. B **109**, 17708 (2005).
160. M. Venturoli, M. M. Sperotto, M. Kranenburg, and B. Smit, Phys. Rep. **437**, 1 (2006).
161. H. Noguchi and M. Takasu, J. Chem. Phys. **115**, 9547 (2001).
162. H. Noguchi, J. Chem. Phys. **117**, 8130 (2002).
163. O. Farago, J. Chem. Phys. **119**, 596 (2003).
164. I. R. Cooke, K. Kremer, and M. Deserno, Phys. Rev. E **72**, 011506 (2005).
165. W. Helfrich and Z. Naturforsch. **28c**, 693 (1973).
166. G. Gompper and D. M. Kroll, J. Phys.: Condens. Matter **9**, 8795 (1997).
167. G. Gompper and D. M. Kroll, in *Statistical Mechanics of Membranes and Surfaces*, edited by D. R. Nelson, T. Piran, and S. Weinberg (World Scientific, Singapore, 2004), pp. 359–426, 2nd ed.
168. J. M. Drouffe, A. C. Maggs, and S. Leibler, Science **254**, 1353 (1991).
169. H. Noguchi and G. Gompper, Phys. Rev. E **73**, 021903 (2006).
170. J.-S. Ho and A. Baumgärtner, Europhys. Lett. **12**, 295 (1990).
171. D. M. Kroll and G. Gompper, Science **255**, 968 (1992).
172. D. H. Boal and M. Rao, Phys. Rev. A **45**, R6947 (1992).
173. G. Gompper and D. M. Kroll, J. Phys. I France **6**, 1305 (1996).
174. C. Itzykson, in *Proceedings of the GIFT Seminar, Jaca 85*, edited by J. Abad, M. Asorey, and A. Cruz (World Scientific, Singapore, 1986), pp. 130–188.
175. G. Gompper, in *Soft Matter – Complex Materials on Mesoscopic Scales*, edited by J. K. G. Dhont, G. Gompper, and D. Richter (Forschungszentrum Jülich, Jülich, 2002), vol. 10 of *Matter and Materials*.
176. U. Seifert, K. Berndl, and R. Lipowsky, Phys. Rev. A **44**, 1182 (1991).
177. G. Gompper and D. M. Kroll, Phys. Rev. Lett. **73**, 2139 (1994).
178. G. Gompper and D. M. Kroll, Phys. Rev. E **51**, 514 (1995).
179. H.-G. Döbereiner, G. Gompper, C. Haluska, D. M. Kroll, P. G. Petrov, and K. A. Riske, Phys. Rev. Lett. **91**, 048301 (2003).
180. H. Noguchi and G. Gompper, Phys. Rev. E **72**, 011901 (2005).
181. B. Alberts, A. Johnson, J. Lewis, M. Raff, K. Roberts, and P. Walter, *Molecular Biology of the Cell* (Garland, New York, 2007), 5th ed.
182. N. Mohandas and E. Evans, Annu. Rev. Biophys. Biomol. Struct. **23**, 787 (1994).
183. L. Scheffer, A. Bitler, E. Ben-Jacob, and F. Korenstein, Eur. Biophys. J. **30**, 83 (2001).
184. G. Lenormand, S. Hénon, A. Richert, J. Siméon, and F. Gallet, Biophys. J. **81**, 43 (2001).
185. D. E. Discher, D. H. Boal, and S. K. Boey, Biophys. J. **75**, 1584 (1998).
186. G. Lim, M. Wortis, and R. Mukhopadhyay, Proc. Natl. Acad. Sci. USA **99**, 16766 (2002).
187. H. Noguchi and G. Gompper, Proc. Natl. Acad. Sci. USA **102**, 14159 (2005).
188. H. Noguchi and G. Gompper, J. Chem. Phys. **125**, 164908 (2006).
189. M. M. Dupin, I. Halliday, C. M. Care, L. Alboul, and L. L. Munn, Phys. Rev. E **75**, 066707 (2007).
190. K. H. de Haas, C. Blom, D. van den Ende, M. H. G. Duits, and J. Mellema, Phys. Rev. E **56**, 7132 (1997).
191. V. Kantsler and V. Steinberg, Phys. Rev. Lett. **95**, 258101 (2005).
192. V. Kantsler and V. Steinberg, Phys. Rev. Lett. **96**, 036001 (2006).
193. M. A. Mader, V. Vitkova, M. Abkarian, A. Viallat, and T. Podgorski, Eur. Phys. J. E **19**, 389 (2006).
194. S. R. Keller and R. Skalak, J. Fluid Mech. **120**, 27 (1982).
195. T. W. Secomb and R. Skalak, Q. J. Mech. Appl. Math. **35**, 233 (1982).
196. T. W. Secomb, T. M. Fischer, and R. Skalak, Biorheology **20**, 283 (1983).
197. R. Tran-Son-Tay, S. P. Sutera, and P. R. Rao, Biophys. J. **46**, 65 (1984).
198. U. Seifert, Eur. Phys. J. B **8**, 405 (1999).
199. C. Misbah, Phys. Rev. Lett. **96**, 028104 (2006).

200. G. Danker, T. Biben, T. Podgorski, C. Verdier, and C. Misbah, Phys. Rev. E **76**, 041905 (2007).
201. V. V. Lebedev, K. S. Turitsyn, and S. S. Vergeles, Phys. Rev. Lett. **99**, 218101 (2007).
202. M. Kraus, W. Wintz, U. Seifert, and R. Lipowsky, Phys. Rev. Lett. **77**, 3685 (1996).
203. J. Beaucourt, F. Rioual, T. Séon, T. Biben, and C. Misbah, Phys. Rev. E **69**, 011906 (2004).
204. T. Biben, K. Kassner, and C. Misbah, Phys. Rev. E **72**, 041921 (2005).
205. H. Noguchi and G. Gompper, Phys. Rev. Lett. **98**, 128103 (2007).
206. F. Brochard and J. F. Lennon, J. Phys. France **36**, 1035 (1975).
207. J. Rudnick and G. Gaspari, J. Phys. A **19**, L191 (1986).
208. G. B. Nash and H. J. Meiselman, Biophys. J. **43**, 63 (1983).
209. K. Tsukada, E. Sekizuka, C. Oshio, and H. Minamitani, Microvasc. Res. **61**, 231 (2001).
210. B. M. Discher, Y.-Y. Won, D. S. Ege, J. C.-M. Lee, F. S. Bates, D. E. Discher, and D. A. Hammer, Science **284**, 1143 (1999).
211. R. Dimova, U. Seifert, B. Pouligny, S. Förster, and H.-G. Döbereiner, Eur. Phys. J. E **7**, 241 (2002).
212. H. Noguchi and G. Gompper, J. Phys.: Condens. Matter **17**, S3439 (2005).
213. R. Finken, A. Lamura, U. Seifert, and G. Gompper, Eur. Phys. J. E **25**, 309 (2008).
214. T. W. Secomb, R. Skalak, N. Özkaya, and J. F. Gross, J. Fluid Mech. **163**, 405 (1986).
215. R. Skalak, Biorheology **27**, 277 (1990).
216. R. Bruinsma, Physica A **234**, 249 (1996).
217. C. Quéguiner and D. Barthès-Biesel, J. Fluid Mech. **348**, 349 (1997).
218. C. Pozrikidis, Phys. Fluids **17**, 031503 (2005).
219. C. Pozrikidis, Ann. Biomed. Eng. **33**, 165 (2005).
220. R. Skalak, Science **164**, 717 (1969).
221. S. Chien, Ann. Rev. Physiol. **49**, 177 (1987).
222. Y. Suzuki, N. Tateishi, M. Soutani, and N. Maeda, Microcirc. **3**, 49 (1996).
223. K. Boryczko, W. Dzwinel, and D. A. Yuen, J. Mol. Modeling **9**, 16 (2003).
224. W. Dzwinel, K. Boryczko, and D. A. Yuen, J. Colloid Int. Sci. **258**, 163 (2003).
225. C. D. Eggleton and A. S. Popel, Phys. Fluids **10**, 1834 (1998).
226. Y. Liu and W. K. Liu, J. Comput. Phys. **220**, 139 (2006).
227. H. Tanaka, J. Phys.: Condens. Matter **12**, R207 (2000).
228. J. Vermant and M. J. Solomon, J. Phys.: Condens. Matter **17**, R187 (2005).
229. Y.-G. Tao, I. O. Götze, and G. Gompper, J. Chem. Phys. **128**, 144902 (2008).
230. R. B. Bird, C. F. Curtis, R. C. Armstrong, and O. Hassager, *Dynamics of Polymeric Liquids*, Volume 2: Kinetic Theory (Wiley, New York, 1987).

Adv Polym Sci 221: 89–166
DOI:10.1007/12_2008_4
© Springer-Verlag Berlin Heidelberg 2008

Lattice Boltzmann Simulations of Soft Matter Systems

Burkhard Dünweg and Anthony J.C. Ladd

Abstract This article concerns numerical simulations of the dynamics of particles immersed in a continuum solvent. As prototypical systems, we consider colloidal dispersions of spherical particles and solutions of uncharged polymers. After a brief explanation of the concept of hydrodynamic interactions, we give a general overview of the various simulation methods that have been developed to cope with the resulting computational problems. We then focus on the approach we have developed, which couples a system of particles to a lattice-Boltzmann model representing the solvent degrees of freedom. The standard D3Q19 lattice-Boltzmann model is derived and explained in depth, followed by a detailed discussion of complementary methods for the coupling of solvent and solute. Colloidal dispersions are best described in terms of extended particles with appropriate boundary conditions at the surfaces, while particles with internal degrees of freedom are easier to simulate as an arrangement of mass points with frictional coupling to the solvent. In both cases, particular care has been taken to simulate thermal fluctuations in a consistent way. The usefulness of this methodology is illustrated by studies from our own research, where the dynamics of colloidal and polymeric systems has been investigated in both equilibrium and nonequilibrium situations.

Keywords Boundary conditions, Brownian motion, Chapman–Enskog, Colloidal dispersions, Fluctuation–dissipation theorem, Force coupling, Hydrodynamic interactions, Hydrodynamic screening, Lattice Boltzmann, Polymer solutions, Soft matter

B. Dünweg (✉)
Max Planck Institute for Polymer Research, Ackermannweg 10, 55128 Mainz, Germany
e-mail: duenweg@mpip-mainz.mpg.de

A.J.C. Ladd
Chemical Engineering Department, University of Florida, Gainesville, FL 32611-6005, USA
e-mail: tladd@che.ufl.edu

Contents

1 Introduction

The term "soft condensed matter" generally refers to materials which possess additional "mesoscopic" length scales between the atomic and the macroscopic scales [1–8]. While simple fluids are characterized by the atomic size (3×10^{-10} m), soft-matter systems contain one or more additional length scales, typically of order 10^{-9}–10^{-6} m. There are many examples of matter with mesoscale structure, including suspensions, gels, foams, and emulsions; all of these are characterized by viscoelastic behavior, which means a response that is fluid-like on long time scales but solid-like on shorter time scales. The two prototype systems considered in this article are colloidal dispersions of hard particles, where the additional length scale is provided by the particle size, and polymer systems, where the length scale is the size of the macromolecule. The main difference between these two systems is the presence of internal degrees of freedom in the polymer, such that a statistical description is necessary even on the single-molecule level. Many additional soft-matter systems exist. For example, dispersions may not only contain spherical particles, but rather rod-like or disk-like objects. For polymers, there are many possible molecular architectures; in addition to simple linear chains, there are rings, stars, combs, bottle-brush polymers and dendrimers. Polymers may also self-assemble into two-dimensional membranes, either free or tethered, which are of paramount biological importance in cell membranes, vesicles, and red blood cells.

Strongly non-linear rheology is characteristic of soft matter. In simple fluids, it is difficult to observe any deviations from Newtonian behavior, which is well described by the hydrodynamic equations of motion with linear transport coefficients that depend only on the thermodynamic state. Indeed, Molecular Dynamics simulations [9] have revealed that a hydrodynamic description is valid down to astonishingly small scales, of the order of a few collisions of an individual molecule. This means that one would have to probe the system with very short wave lengths and very high frequencies, which are typically not accessible to standard experiments (with the exception of neutron scattering [10]), and even less in everyday life. However, in soft-matter systems microstructural components (particles and polymers for example) induce responses that depend very much on frequency and length scale. These systems are often referred to as "complex fluids."

The nonlinear rheological properties of soft matter pose a substantial challenge for theory [11]. Therefore, the study of simple model systems is often the only way to make systematic progress. Numerical simulations allow us to follow the dynamics of model systems without invoking the uncontrolled approximations that are usually required by purely analytical methods. Simulations can be used to isolate and investigate the influence of microstructure, composition, external perturbation and geometry in ways that cannot always be duplicated in the laboratory. In particular, simulations can provide a well-defined test bed for theoretical ideas, allowing them to be evaluated in a simpler and more rigorous environment than is possible experimentally. Finally, they can provide more detailed and direct information on the particle dynamics and structure than is typically possible with experimental measurements.

In this article we will focus on systems which comprise particles, with or without internal degrees of freedom, suspended in a simple fluid. We will first outline the necessary ingredients for a theoretical description of the dynamics, and in particular explain the concept of hydrodynamic interactions (HI). Starting from this background, we will provide a brief overview of the various simulation approaches that have been developed to treat such systems. All of these methods are based upon a description of the solute in terms of particles, while the solvent is taken into account by a simple (but sufficient) model, making use of the fact that it can be described as a Newtonian fluid. Such methods are often referred to as "mesoscopic." We will then describe and derive in some detail the algorithms that have been developed by us to couple a particulate system to a LB fluid. The usefulness of these methods will then be demonstrated by applications to colloidal dispersions and polymer solutions. Some of the material presented here is a summary of previously published work.

2 Particle-Fluid Systems

2.1 Coarse-Grained Models

The first step towards understanding systems of particles suspended in a solvent is the notion of *scale separation*. Colloidal particles are much larger (up to 10^{-6} m) than solvent molecules, and their relaxation time (up to 10 s), given by the time the particle needs to diffuse its own size, is several orders of magnitude larger than the corresponding solvent time scale (10^{-12} s) as well. A similar separation of scales holds for solute with internal degrees of freedom, like polymer chains, membranes, or vesicles, as far as their diffusive motion or their global conformational reorganization is concerned. However, these systems contain a hierarchy of length and time scales, which can be viewed as a spectrum of internal modes with a given wavelength and relaxation time. For the long-wavelength part of this spectrum, the same scale separation holds, but when the wavelength is comparable with the solvent size molecular interactions become important. However, on scales of interest, these details can usually be lumped into a few parameters. The solute is then modeled as a system of "beads" interacting with each other via an effective potential. The beads should be viewed as collections of atomic-scale constituents, either individual atoms or functional groups, which have been combined into a single effective unit in a process known as "coarse-graining." On the scale of the beads, the solvent may be viewed as a hydrodynamic continuum, characterized by its shear viscosity and temperature. The flow of soft matter is usually isothermal, incompressible, and inertia-free (zero Reynolds number). Then, the most natural parameter to describe solute–solvent coupling in polymeric systems is the Stokes friction coefficient of the individual beads, or (in the case of anisotropic subunits) the corresponding tensorial generalization. The effective friction coefficient is the lumped result of a more detailed, or "fine-grained," description at the molecular scale, as is the bead–bead potential, which should be viewed as a potential of mean force. In order to make contact with experimental systems, these parameters must be calculated from more microscopic theories or simulations, or deduced from experimental data.

Further simplification may arise from the type of scientific question being addressed by the coarse-grained model. If the interest is not in specific material properties of a given chemical species, but rather in generic behavior and mechanisms, as is the case for the examples we will discuss in this article, then the details of the parameterization are less important than a model that is both conceptually simple and computationally efficient. The standard model for particulate suspensions is a system of hard spheres, while for polymer chains the Kremer–Grest model [12] has proved to be a valuable and versatile tool. Here, the beads interact via a purely repulsive Lennard-Jones (or WCA [13]) potential,

$$V_{LJ}(r) = \begin{cases} 4\varepsilon \left[\left(\frac{\sigma}{r}\right)^{12} - \left(\frac{\sigma}{r}\right)^{6} + \frac{1}{4} \right] & r \leq 2^{1/6}\sigma, \\ 0 & r \geq 2^{1/6}\sigma, \end{cases} \qquad (1)$$

while consecutive beads along the chain are connected via a FENE ("finitely extensible nonlinear elastic") potential,

$$V_{\mathrm{ch}}(r) = -\frac{k}{2}R_0^2 \ln\left(1 - \frac{r^2}{R_0^2}\right), \tag{2}$$

with typical parameters $R_0/\sigma = 1.5$, $k\sigma^2/\varepsilon = 30$. Chain stiffness can be incorporated by an additional bond-bending potential, while more complicated architectures (like stars and tethered membranes) require additional connectivity. Poor solvent quality is modeled by adding an attractive part to the non-bonded interaction, while Coulomb interactions (however without local solvent polarization) can be included by using charged beads.

There are two further ingredients which any good soft-matter model should take into account: On the one hand, *thermal fluctuations* are needed in order to drive Brownian motion and internal reorganization of the conformational degrees of freedom. There are, however, special situations where thermal noise can be disregarded, and ideally the simulation method should be flexible enough to be able to turn the noise both on and off. On the other hand, *hydrodynamic interactions*, which will be the subject of the next subsection, need to be taken into account in most circumstances.

2.2 Hydrodynamic Interactions

The term "hydrodynamic interactions" describes the *dynamic* correlations between the particles, induced by diffusive momentum transport through the solvent. The physical picture is the same, whether the particle motion is Brownian (i.e., driven by thermal noise) or the result of an external force (e.g., sedimentation or electrophoresis). The motion of particle i perturbs the surrounding solvent, and generates a flow. This signal spreads out diffusively, at a rate governed by the kinematic viscosity of the fluid $\eta_{\mathrm{kin}} = \eta/\rho$ (η is the solvent shear viscosity and ρ is its mass density). On interesting (long) time scales, only the transverse hydrodynamic modes [14] remain, and the fluid may be considered as incompressible. The viscous momentum field around a particle diffuses much faster than the particle itself, so that the Schmidt number

$$Sc = \frac{\eta_{\mathrm{kin}}}{D} \tag{3}$$

is large. In a molecular fluid, D is the diffusion coefficient of the solvent molecules and $Sc \sim 10^2$–10^3, while for soft matter, where D is the diffusion coefficient of the polymer or colloid, Sc is much larger, up to 10^6 for micrometer size colloids. The condition $Sc \gg 1$ is an important restriction on the dynamics, which good mesoscopic models should satisfy. Consider two beads i and j, separated by a distance r_{ij}. The momentum generated by the motion of i with respect to the surrounding fluid reaches bead j after a "retardation time" $\tau \sim r_{ij}^2/\eta_{\mathrm{kin}}$. After this time the motion of

j becomes *correlated* with that of bead i. However, during the retardation time the beads have traveled a small distance $\sim\sqrt{D\tau} \sim r_{ij}/\sqrt{Sc}$, which is negligible in comparison with r_{ij}. Therefore, it is quite reasonable to describe the Brownian motion of the beads neglecting retardation effects, and consider their random displacements to be *instantaneously* correlated.

These general considerations suggest a Langevin description (stochastic differential equation) for the time evolution of the bead positions $\mathbf{r}_i$:

$$\frac{d}{dt}r_{i\alpha} = \sum_j \mu_{i\alpha,j\beta} F^c_{j\beta} + \Delta_{i\alpha}, \tag{4}$$

where α, β are Cartesian indexes and the Einstein summation convention has been assumed. $\mathbf{F}^c_j$ is the conservative force acting on the jth bead, while the mobility tensor μ_{ij} describes the velocity response of particle i to $\mathbf{F}^c_j$; both $\mathbf{F}^c_j$ and μ_{ij} depend on the configuration of the N particles, $\mathbf{r}^N$. The random displacements (per unit time) $\Delta_{i\alpha}$ are Gaussian white noise variables [15,16] satisfying the fluctuation–dissipation relation

$$\langle\Delta_{i\alpha}\rangle = 0, \tag{5}$$

$$\langle\Delta_{i\alpha}(t)\Delta_{j\beta}(t')\rangle = 2k_\mathrm{B}T\mu_{i\alpha,j\beta}\delta(t-t'); \tag{6}$$

here T is the temperature and k_B denotes the Boltzmann constant. In general, the mobility tensor, $\mu_{i\alpha,j\beta}$ depends on particle position, in which case the stochastic integral should be evaluated according to the Stratonovich calculus. This is because during a finite time step a particle samples different mobilities as its position changes. Numerically simpler is the Ito calculus [15, 16], which uses the mobility at the beginning of the time step. In this case an additional term needs to be added to (4),

$$\frac{d}{dt}r_{i\alpha} = \sum_j \mu_{i\alpha,j\beta} F^c_{j\beta} + k_\mathrm{B}T \sum_j \frac{\partial\mu_{i\alpha,j\beta}}{\partial r_{j\beta}} + \Delta_{i\alpha}. \tag{7}$$

In cases where the divergence of the mobility vanishes, $\partial\mu_{i\alpha,j\beta}/\partial r_{j\beta} = 0$, the Ito and Stratonovich interpretations coincide.

Processes generated by (4) (Stratonovich) or (7) (Ito) sample trajectories from a probability distribution in the configuration space of the beads, $P(\mathbf{r}^N,t)$, which evolves according to a Fokker–Planck equation (Kirkwood diffusion equation):

$$\frac{\partial}{\partial t}P(\mathbf{r}^N,t) = \mathcal{L}P(\mathbf{r}^N,t), \tag{8}$$

with the Fokker–Planck operator

$$\mathcal{L} = \sum_{ij} \frac{\partial}{\partial r_{i\alpha}}\mu_{i\alpha,j\beta}\left(k_\mathrm{B}T\frac{\partial}{\partial r_{j\beta}} - F^c_{j\beta}\right). \tag{9}$$

Writing the forces as the derivative of a potential,

$$F_{i\alpha}^{c} = -\frac{\partial}{\partial r_{i\alpha}} V\left(\mathbf{r}^{N}\right), \tag{10}$$

it follows that the Boltzmann distribution, $P \propto \exp\left(-V/k_{B}T\right)$, is the stationary solution of the Kirkwood diffusion equation; in other words the model satisfies the fluctuation–dissipation relation.

The mobility tensor can be derived from Stokes-flow hydrodynamics. Consider a set of spherical particles, located at positions $\mathbf{r}_{i}$ with radius a, surrounded by a fluid with shear viscosity η. Each of the particles has a velocity $\mathbf{v}_{i}$, which, as a result of stick boundary conditions, is identical to the local fluid velocity on the particle surface. The resulting fluid motions generate hydrodynamic drag forces $\mathbf{F}_{i}^{d}$, which at steady state are balanced by the conservative forces, $\mathbf{F}_{i}^{d} + \mathbf{F}_{i}^{c} = 0$. The commonly used approximation scheme is a systematic multipole expansion, similar to the analogous expansion in electrostatics [17–21]. For details, we refer the reader to the original literature [17], where the contributions from rotational motion of the beads are also considered. As a result of the linearity of Stokes flow, the particle velocities and drag forces are linearly related,

$$v_{i\alpha} = -\sum_{j}\mu_{i\alpha,j\beta}F_{j\beta}^{d} = \sum_{j}\mu_{i\alpha,j\beta}F_{j\beta}^{c}. \tag{11}$$

Since μ_{ij} describes the velocity response of particle i to the force acting on particle j, it must be identical to the mobility tensor appearing in the Langevin equation.

In general the mobility matrix is a function of all the particle coordinates, but to leading order, it is pairwise additive:

$$\mu_{ij} = \frac{\delta_{ij}\mathbf{1}}{6\pi\eta a} + \frac{(1-\delta_{ij})}{8\pi\eta r_{ij}}\left(1 + \frac{\mathbf{r}_{ij}\mathbf{r}_{ij}}{r_{ij}^{2}}\right) + \frac{(1-\delta_{ij})a^{2}}{12\pi\eta r_{ij}^{3}}\left(1 - 3\frac{\mathbf{r}_{ij}\mathbf{r}_{ij}}{r_{ij}^{2}}\right), \tag{12}$$

where $\mathbf{r}_{ij} = \mathbf{r}_{i} - \mathbf{r}_{j}$, and $\mathbf{1}$ denotes the unit tensor. The hydrodynamic interaction is long-ranged and therefore has a strong influence on the collective dynamics of suspensions and polymer solutions. This approximate form for the mobility matrix follows from the assumption that the force density on the sphere surface is constant. A point multipole expansion, by contrast, generates the Oseen ($1/r_{ij}$) interaction at lowest order [22], and can lead to non-positive-definite mobility matrices [23]. Thus, the simplest practical form for the hydrodynamic interaction is the Rotne–Prager tensor [23] given in (12). Both the Oseen and Rotne–Prager mobilities are divergence free, and therefore there is no distinction between Ito and Stratonovich interpretations. However, at higher orders in the multipole expansion, the divergence is non-zero [24].

2.3 Computer Simulation Methods and Models

In this section we briefly summarize the Brownian dynamics algorithm and its close cousin Stokesian Dynamics. We then outline the motivation and development of several mesoscale methods, some of which are reviewed elsewhere in this series.

2.3.1 Brownian Dynamics

Brownian dynamics is conceptually the most straightforward approach [25, 26]. Starting from the Langevin equation for the particle coordinates, (4), and discretizing the time into finite length steps h, gives a first-order (Euler) update for the particle positions,

$$r_{i\alpha}(t+h) = r_{i\alpha}(t) + \sum_j \mu_{i\alpha,j\beta} F^c_{j\beta} h + \sqrt{2k_B T h} \sum_j \sigma_{i\alpha,j\beta} q_{j\beta}. \qquad (13)$$

Here $q_{i\alpha}$ are random variables with

$$\langle q_{i\alpha} \rangle = 0, \qquad (14)$$
$$\langle q_{i\alpha} q_{j\beta} \rangle = \delta_{ij}\delta_{\alpha\beta}, \qquad (15)$$

while the matrix $\sigma_{i\alpha,j\beta}$ satisfies the relation

$$\sum_k \sigma_{i\alpha,k\gamma}\sigma_{j\beta,k\gamma} = \mu_{i\alpha,j\beta}. \qquad (16)$$

Note that in (13) we have assumed a divergence-free mobility tensor

Although the number of degrees of freedom has been minimized, this approach is computationally intensive, and imposes severe limitations on the size of the system that can be studied. Since every particle interacts with every other particle, the calculation of the mobility matrix scales as $O(N^2)$, where N is the number of Brownian particles. In addition, the covariance matrix for the random displacements requires a Cholesky decomposition of the mobility matrix, which scales as $O(N^3)$ [27]. The computational costs of Brownian dynamics are so large that even today one cannot treat more than a few hundred Brownian particles [28].

"Stokesian Dynamics" [29] is an improved version of Brownian dynamics, in which the mobility tensor takes into account short-range (lubrication) contributions to the hydrodynamic forces. It also improves the far-field interactions by including contributions from torques and stresslets, although still higher moments are needed for accurate results in concentrated suspensions [19]. Stokesian Dynamics is even more computationally intensive than Brownian dynamics; the determination of the mobility tensor is already an $O(N^3)$ process.

However, there have been two important improvements in efficiency. First, Fixman [25, 26, 30, 31] has proposed an approximation to $\sigma_{i\alpha,j\beta}$ by a truncated expansion in Chebyshev polynomials, which has a more favorable scaling than Cholesky decomposition. Second, the long-range hydrodynamic interactions can be calculated by Fast Fourier Transforms [32–37], or hierarchical multipole expansions [20]. Accelerated Brownian Dynamics and Stokesian Dynamics algorithms scale close to linearly in the number of particles, and their full potential is not yet explored. However, it should be noted that all these methods are based upon an efficient evaluation of the Green's function for the Stokes flow, which depends on the global boundary conditions. For planar boundaries, solutions are available [21, 38–40], but a more general shape requires a numerical calculation of the Green's function between a tabulated set of source and receiver positions [41].

2.3.2 Mesoscale Methods

In view of the computational difficulties associated with Brownian Dynamics, several "mesoscale" methods have been developed recently. The central idea is to *keep* the solvent degrees of freedom, but to describe them in a simplified fashion, such that only the most salient features survive. As we have already seen, it is in principle sufficient to describe the solvent as a Navier–Stokes continuum, *or* by some suitable model which behaves like a Navier–Stokes continuum on sufficiently large length and time scales. At least asymptotically, the solvent dynamics must be described by the equations

$$\partial_t \rho + \partial_\alpha (\rho u_\alpha) = 0,$$
$$\partial_t (\rho u_\alpha) + \partial_\beta (\rho u_\alpha u_\beta) + \partial_\alpha p = \partial_\beta \sigma_{\alpha\beta} + \partial_\beta \sigma^f_{\alpha\beta} + f_\alpha, \qquad (17)$$

where ρ is the mass density, $\rho\mathbf{u}$ the momentum density, p the thermodynamic pressure, $\mathbf{f}$ an external force density applied to the fluid, σ the viscous stress tensor, and σ^f the fluctuating (Langevin) stress [42], whose statistical properties will be discussed in later sections of this article. The viscous stresses are characterized by the shear and bulk viscosities, η and η_v, which we will assume to be constants, independent of thermodynamic state and flow conditions:

$$\sigma_{\alpha\beta} = \eta \left(\partial_\alpha u_\beta + \partial_\beta u_\alpha - \frac{2}{3} \partial_\gamma u_\gamma \delta_{\alpha\beta} \right) + \eta_v \partial_\gamma u_\gamma \delta_{\alpha\beta}. \qquad (18)$$

The advantage of such approaches is their spatial locality, resulting in favorable $O(N)$ scaling, combined with ease of implementation and parallelization. The disadvantage is the introduction of additional degrees of freedom, and of additional (short) time scales which are not of direct interest. The coupling between solvent and solute varies from method to method. However, in all cases one takes the masses and the momenta of the solute particles explicitly into account, and makes sure that the total momentum is conserved.

Lattice models (Navier–Stokes, lattice Boltzmann) simulate a discretized field theory in which thermal fluctuations can be added, but also avoided if desired.

Particle methods (Molecular Dynamics, Dissipative Particle Dynamics, Multi-Particle Collision Dynamics) simulate a system of interacting mass points, and therefore thermal fluctuations are always present. The particles may have size and structure or they may be just point particles. In the former case, the finite solvent size results in an additional potential of mean force between the beads. The solvent structure extends over unphysically large length scales, because the proper separation of scale between solute and solvent is not computationally realizable. In dynamic simulations of systems in thermal equilibrium [43], solvent structure requires that the system be equilibrated with the solvent in place, whereas for a structureless solvent the solute system can be equilibrated by itself, with substantial computational savings [43]. Finally, lattice models have a (rigorously) known solvent viscosity, whereas for particle methods the existing analytical expressions are only approximations (which however usually work quite well).

These considerations suggest that lattice methods are somewhat more flexible and versatile for soft-matter simulations. On the other hand, the coupling between solvent and immersed particles is less straightforward than for a pure particle system. The coupling between solid particles and a lattice-based fluid model will be discussed in detail in Sect. 4.

2.3.3   Molecular Dynamics

Molecular Dynamics (MD) is the most fundamental approach to soft-matter simulations. Here the solute particles are immersed in a bath of solvent molecules and Newton's equations of motion are solved numerically. In this case, it is impossible to make the solvent structureless – a structureless solvent would be an ideal gas of point particles, which never reaches thermal equilibrium. Furthermore, the model interaction potentials are stiff and considerable simulation time is spent following the motion of the solvent particles in their local "cages." These disadvantages are so severe that nowadays MD is rarely applied to soft-matter systems of the type we are discussing in this article.

2.3.4   Dissipative Particle Dynamics

Dissipative Particle Dynamics (DPD), which has become quite popular in the soft-matter community [44–56], was developed to address the computational limitations of MD. A very soft interparticle potential, representing coarse-grained aggregates of molecules, enables a large time step to be used. Furthermore, a momentum-conserving Galilean-invariant thermostat is included, representing the degrees of freedom that have been lost in the coarse-graining process. Practically, these two parts are unrelated, such that it is legitimate to apply the DPD thermostat to a standard MD system. The DPD thermostat is consistent with macroscopic isothermal thermodynamics. Since this already introduces interparticle collisions, it is possible to run DPD using an ideal gas solvent and still achieve thermal equilibrium.

The key innovation in DPD is to apply the thermostat to particle *pairs*. A frictional damping is applied to the *relative* velocities between each neighboring pair, and a corresponding random force is added in a pairwise fashion also, such that Newton's third law holds exactly. The implementation is as follows. We define two functions $\zeta(r) \geq 0$, the relative friction coefficient for particle pairs with interparticle distance r, and $\sigma(r) \geq 0$, characterizing the strength of the stochastic force applied to the same particle pair. The fluctuation–dissipation theorem requires that

$$\sigma^2(r) = k_{\mathrm{B}} T \zeta(r). \tag{19}$$

The functions have compact support, so that only near neighbors need be taken into account.

The frictional force on particle i is determined by projecting the relative velocities onto the interparticle separation ($\hat{r}_{ij} = \mathbf{r}_{ij}/|\mathbf{r}_{ij}|$):

$$\mathbf{F}_i^{\mathrm{d}} = -\sum_j \zeta(r_{ij}) \left[(\mathbf{v}_i - \mathbf{v}_j) \cdot \hat{r}_{ij} \right] \hat{r}_{ij}, \tag{20}$$

which conserves momentum exactly, $\sum_i \mathbf{F}_i^{\mathrm{d}} = 0$. Similarly, the stochastic forces are directed along the interparticle separation, again so that momentum is conserved pair-by-pair,

$$\mathbf{F}_i^{\mathrm{f}} = \sum_j \sigma(r_{ij}) \, \eta_{ij}(t) \, \hat{r}_{ij}. \tag{21}$$

The noise η_{ij} satisfies the relations $\eta_{ij} = \eta_{ji}$, $\langle \eta_{ij} \rangle = 0$, and

$$\langle \eta_{ij}(t) \eta_{kl}(t') \rangle = 2(\delta_{ik}\delta_{jl} + \delta_{il}\delta_{jk}) \delta(t - t'), \tag{22}$$

such that different pairs are statistically independent and $\sum_i \mathbf{F}_i^{\mathrm{f}} = 0$. The equations of motion for a particle of mass m_i and momentum $\mathbf{p}_i$ are

$$\frac{\mathrm{d}}{\mathrm{d}t} \mathbf{r}_i = \frac{1}{m_i} \mathbf{p}_i, \tag{23}$$

$$\frac{\mathrm{d}}{\mathrm{d}t} \mathbf{p}_i = \mathbf{F}_i^{\mathrm{c}} + \mathbf{F}_i^{\mathrm{d}} + \mathbf{F}_i^{\mathrm{f}}. \tag{24}$$

Exploiting the relation between this stochastic differential equation and its Fokker–Planck equation, it can be shown that the fluctuation–dissipation theorem holds [46], and that the method therefore simulates a canonical ensemble. DPD can be extended to thermalize the perpendicular component of the interparticle velocity as well, thereby allowing more control over the transport properties of the model [49, 57].

2.3.5 Multi-Particle Collision Dynamics

This method [58–62] works with a system of ideal-gas particles and therefore has
no artificial depletion forces. Free streaming of the particles,

$$\mathbf{r}_i(t+h) = \mathbf{r}_i(t) + h\mathbf{v}_i(t), \tag{25}$$

alternates with momentum and energy conserving collisions, which are implemented via a Monte Carlo procedure:

- Sub-divide the simulation volume into a regular array of cells.
- For each cell, determine the set of particles residing in it. For one particular cell,
  let these particles be numbered $i = 1,\ldots,n$. Then in each box:
- Determine the local center-of-mass velocity:

$$\mathbf{v}_{\mathrm{CM}} = \frac{1}{n}\sum_{i=1}^{n}\mathbf{v}_i. \tag{26}$$

- For each particle in the cell, perform a Galilean transformation into the local
  center-of-mass system:

$$\tilde{\mathbf{v}}_i = \mathbf{v}_i - \mathbf{v}_{\mathrm{CM}}. \tag{27}$$

- Within the local center-of-mass system, rotate all velocities within the cell by a
  random rotation matrix R:

$$\tilde{\mathbf{v}}_i' = \mathsf{R}\tilde{\mathbf{v}}_i. \tag{28}$$

- Transform back into the laboratory system:

$$\mathbf{v}_i' = \tilde{\mathbf{v}}_i' + \mathbf{v}_{\mathrm{CM}}. \tag{29}$$

By suitable random shifts of the cells relative to the fluid, it is possible to recover
strict Galilean invariance [59, 60]. Multi-Particle Collision Dynamics (MPCD) results in hydrodynamic behavior on large length and time scales, and is probably the
simplest and most efficient particle method to achieve this.

2.3.6 Lattice Boltzmann

Here one solves the Boltzmann equation, known from the kinetic theory of gases, in
a fully discretized fashion. Space is discretized into a regular array of lattice sites,
time is discretized, and velocities are chosen such that one time step will connect
only nearby lattice sites. Free streaming along the lattice links alternates with local
on-site collisions. Care must be taken to restore isotropy and Galilean invariance
in the hydrodynamic limit, and asymptotic analysis is an indispensable tool in this
process. Further details will be provided in the following sections.

2.3.7 Navier–Stokes

It is possible to start from a discrete representation of (17) but this has not been particularly popular in soft-matter simulations, due to the difficulty of including thermal fluctuations (but see [63]). Finite-difference methods share many technical similarities with lattice Boltzmann (LB) and are roughly comparable in terms of computational resources. However, to our knowledge, no detailed benchmark comparisons are available as yet. In order to be competitive with LB, we believe that the solver must (1) make sure that mass and momentum are conserved within machine accuracy, as is the case for LB, and (2) *not* work in the incompressible limit, in order to avoid the costly non-local constraints imposed by the typical Poisson solver for the pressure. The incompressible limit is an approximation, which eliminates the short time scales associated with wave-like motion. However, in soft matter the solute particles must be simulated on short inertial time scales, which requires that the solvent is simulated on rather short time scales as well. For this reason, we believe that enforcing an incompressibility constraint does not pose a real advantage, and it is instead preferable to allow for finite compressibility, such that one obtains an explicit and local algorithm. This idea is analogous to the Car–Parrinello method [64], where the Born–Oppenheimer constraint is also discarded, in favor of an approximate but adequate separation of time scales. For simulations of soft-matter systems coupled to a Navier–Stokes background, see [65–74].

3 The Fluctuating Lattice-Boltzmann Equation

The motivation for the development of lattice gas cellular automata (LGCA) [75,76] was to apply a highly simplified MD to simulations of hydrodynamic flows. In LGCA, particles move along the links of a regular lattice, typically cubic or triangular. Each lattice direction is encoded with a label i and a vector $h\mathbf{c}_i$ connects neighboring pairs of sites. During each time step h, all particles with a direction i are displaced $h\mathbf{c}_i$ to an adjacent lattice site; thus $\mathbf{c}_i$ is the (constant) velocity of particles of type i. Interparticle interactions are reduced to collisions between particles on the same lattice site, such that the conservation laws for mass and momentum are satisfied; in single speed LGCA models [75,76], mass conservation implies energy conservation as well. The LB method [77–80] was developed to reduce the thermal noise in LGCA, which requires extensive averaging to obtain statistically significant results.

The LB model preserves the structural simplicity of LGCA, but substitutes an ensemble-averaged collision operator for the detailed microscopic dynamics of the LGCA. The hydrodynamic flow fields develop without thermal noise, but the underlying connection with statistical mechanics is lost (Sect. 3.1). The LB model turns out to be more flexible than LGCA, and there is now a rich literature that includes thermal [81–86] and multiphase flows, involving both liquid–gas coexistence and multicomponent mixtures [87–96]. In the present article, we will consider only single-phase flows of a single solvent species, such that we can describe the dynamics in terms of a single particle type. The algorithm can be summarized by the equation

$$v_i(\mathbf{r} + \mathbf{c}_i h, t + h) = v_i^\star(\mathbf{r}, t) = v_i(\mathbf{r}, t) + \Delta_i(v(\mathbf{r}, t)), \tag{30}$$

where $v_i(\mathbf{r}, t)$ is the number of particles that, at the discrete time t just prior to collision, reside at the lattice site $\mathbf{r}$, and have velocity $\mathbf{c}_i$; $v_i^\star(\mathbf{r}, t)$ indicates the velocity distribution immediately after collision. The difference Δ_i between the pre- and post-collision states is called the "collision operator" and depends on the complete set of populations at the site $v(\mathbf{r}, t)$. The left-hand side of (30) describes the advection of the populations along the links connecting neighboring lattice sites. The velocity set $\mathbf{c}_i$ is chosen such that each new position $\mathbf{r} + \mathbf{c}_i h$ is again at a lattice site; $\mathbf{c}_i = 0$ is possible.

For simplicity and computational efficiency, the number of velocities should be small. Therefore the set of velocities, $\mathbf{c}_i$, is typically limited to two or three neighbor shells, chosen to be compatible with the symmetry of the lattice. In two dimensions a single shell of six neighbors is sufficient for hydrodynamic flows, but a single set of cubic lattice vectors leads to anisotropic momentum diffusion, even at large spatial scales. Thus, LB models employ a judicious mixture of neighboring shells, suitably weighted so that isotropy is recovered. We use the classification scheme introduced by Qian et al. [97]: for instance D2Q9 refers to an LB model on a square lattice in two dimensions, using nine velocities (zero, four nearest neighbors, four next-nearest neighbors), while D3Q19 indicates a three-dimensional model on a simple cubic lattice with 19 velocities (zero, six nearest neighbors, 12 next-nearest neighbors).

3.1 Fluctuations

The difference between lattice gas and LB lies in the nature of the v_i. In a lattice gas v_i is a Boolean variable (i.e., only the values zero and one are allowed), while in the LB equation it is a positive real-valued variable. In Sect. 3.6 we will consider the case where v_i is a large positive integer, a conceptual model we call a "Generalized Lattice Gas" (GLG). Thinking of these models as a simplified MD, and considering fluctuations in v_i, it becomes clear what the key difference between LGCA and the LB equation is. We define a dimensionless "Boltzmann number," Bo, by the fluctuations in v_i at a single site,

$$Bo = \frac{\left(\langle v_i^2 \rangle - \langle v_i \rangle^2\right)^{1/2}}{\langle v_i \rangle}, \tag{31}$$

where $\langle\ldots\rangle$ denotes the ensemble average. One could define Boltzmann numbers for other observables, but they would all produce similar values. The important point is that Bo tells us how coarse-grained the model is, compared to microscopic MD: $Bo \sim 1$ (the maximum value) corresponds to a fully microscopic model where fluctuations are of the same order as the mean. This is exactly the case for LGCA, which should therefore be viewed as a simplified, but not coarse-grained MD. Conversely, deterministic LB algorithms, at sufficiently small Reynolds numbers, and with time-independent driving forces, bring the system to a stationary state with well-defined values for the v_i. In other words, they are characterized by $Bo = 0$, which is the minimal value, corresponding to entirely deterministic physics.

Originally, LGCA and LB algorithms were developed to simulate macroscopic hydrodynamics. Here, a large Boltzmann number (order 1) is undesirable, since the hydrodynamic behavior is only revealed after extensive sampling. For many macroscopic applications a deterministic LB simulation at $Bo = 0$ is hence entirely appropriate. In reality, however, the Boltzmann number is finite, since the spatial domain in the physical system corresponding to a single lattice site is also finite. In soft-matter applications the spatial scales are so small that these fluctuations do need to be taken into account, although in many cases Bo is fairly small. This suggests it would be advantageous to introduce small thermal fluctuations into the LB algorithm, in a controlled fashion, by means of a *stochastic* collision operator [98–100]. The fluctuation–dissipation relation can be satisfied by enforcing consistency with fluctuating hydrodynamics [42] on large length and time scales. An important refinement is to thermalize the additional degrees of freedom that are not directly related to hydrodynamics [101], which leads to equipartition of fluctuation energy on all length scales. A comprehensive understanding of these approaches in terms of the statistical mechanics of LB systems has been achieved only recently [102].

The number variables, v_i, can be connected to the hydrodynamic fields, mass density $\rho(\mathbf{r},t)$, momentum density $\mathbf{j}(\mathbf{r},t)$, and fluid velocity $\mathbf{u}(\mathbf{r},t)$ ($\mathbf{j} = \rho\mathbf{u}$), by introducing the mass of an LB particle, m_p, and the mass density parameter

$$\mu = \frac{m_\mathrm{p}}{b^3}; \tag{32}$$

here b is the lattice spacing and a three-dimensional lattice has been assumed. We then use the mass densities of the individual populations,

$$n_i(\mathbf{r},t) = \mu v_i(\mathbf{r},t), \tag{33}$$

to re-write the LB equation as

$$n_i(\mathbf{r}+\mathbf{c}_i h,t+h) = n_i^\star(\mathbf{r},t) = n_i(\mathbf{r},t) + \Delta_i(\mathbf{n}(\mathbf{r},t)). \tag{34}$$

The mass and momentum densities, ρ and $\mathbf{j}$, are moments of the n_i's with respect to the velocity vectors,

$$\rho(\mathbf{r},t) = \sum_i n_i(\mathbf{r},t), \tag{35}$$

$$\mathbf{j}(\mathbf{r},t) = \sum_i n_i(\mathbf{r},t)\mathbf{c}_i, \tag{36}$$

and therefore, the collision operator must satisfy the constraints of mass and momentum conservation,

$$\sum_i \Delta_i = \sum_i \Delta_i \mathbf{c}_i = 0. \tag{37}$$

The LB algorithm has both locality and conservation laws built in, but two important symmetries have been lost. The system will in general exhibit cubic anisotropy, due to the underlying lattice symmetries, and violate Galilean invariance, due to the finite number of velocities. Isotropy can be restored in the large-scale limit by a careful choice of velocities and collision operator; however, the broken Galilean invariance restricts the method to flows with $u \ll c_i$. Since the speed of sound c_s, the maximum velocity with which any signal can travel through the system, is of the order of the c_i, the condition actually means low Mach number (Ma) flow,

$$Ma = u/c_s \ll 1. \tag{38}$$

In soft-matter applications, variations in fluid density are small and there is a universal equation of state characterized by the pressure at the mean fluid density and temperature, $p_0 = p(\rho_0, T)$, and the speed of sound $c_s = (\partial p/\partial \rho)^{1/2}$ [42],

$$p = p_0 + (\rho - \rho_0)c_s^2. \tag{39}$$

Within an unimportant constant $(p_0 - \rho_0 c_s^2)$, (39) can be replaced by the relation

$$p = \rho c_s^2, \tag{40}$$

which fits well to the linear structure of (34). The value of c_s is immaterial except that it establishes a time-scale separation between sound propagation and viscous diffusion of momentum. For this reason, a model where c_s is unphysically small may be used, so long as the dimensionless number $C_\eta = \rho c_s l / \eta$ is sufficiently large; here l is a characteristic length in the system and η is the shear viscosity of the fluid. For polymers and colloids, $C_\eta \sim 10$–$1{,}000$, but values of C_η in excess of 10 lead to quantitatively similar results.

The simplest equation of state of the form of (40) is an ideal gas,

$$p = \frac{\rho}{m_p} k_B T, \tag{41}$$

where T is the absolute temperature and k_B Boltzmann's constant. Comparison with (40) yields

$$k_B T = m_p c_s^2. \tag{42}$$

The temperature is then determined by choosing values for the discretization parameters b and h ($c_s \sim b/h$), and the LB particle mass m_p. The parameter m_p controls the noise level in stochastic LB simulations [102]: the smaller m_p (at fixed c_s), the smaller the temperature (or the noise level). This makes physical sense, since small m_p means that a fixed amount of mass ρb^3 is distributed onto many particles, and therefore the fluctuations are small.

In this section we will study the connection between the LB equation, (34), and the equations of fluctuating hydrodynamics [42],

$$\partial_t \rho + \partial_\alpha j_\alpha = 0, \qquad (43)$$
$$\partial_t j_\alpha + \partial_\beta \left(\rho c_s^2 \delta_{\alpha\beta} + \rho u_\alpha u_\beta \right) = \partial_\beta \sigma_{\alpha\beta} + \partial_\beta \sigma_{\alpha\beta}^f. \qquad (44)$$

The Greek indexes denote Cartesian components, $\delta_{\alpha\beta}$ is the Kronecker delta, and the Einstein summation convention is implied. The viscous stress has a Newtonian constitutive law,

$$\sigma_{\alpha\beta} = \eta_{\alpha\beta\gamma\delta} \partial_\gamma u_\delta, \qquad (45)$$

and for an isotropic fluid

$$\eta_{\alpha\beta\gamma\delta} = \eta \left(\delta_{\alpha\gamma}\delta_{\beta\delta} + \delta_{\alpha\delta}\delta_{\beta\gamma} - \frac{2}{3}\delta_{\alpha\beta}\delta_{\gamma\delta} \right) + \eta_v \delta_{\alpha\beta}\delta_{\gamma\delta}, \qquad (46)$$

with shear and bulk viscosities η and η_v. The fluctuating stress tensor, $\sigma_{\alpha\beta}^f$, is a Gaussian random variable characterized by zero mean, $\left\langle \sigma_{\alpha\beta}^f \right\rangle = 0$, and a covariance matrix

$$\left\langle \sigma_{\alpha\beta}^f (\mathbf{r},t) \sigma_{\gamma\delta}^f (\mathbf{r}',t') \right\rangle = 2k_B T \eta_{\alpha\beta\gamma\delta} \delta (\mathbf{r} - \mathbf{r}') \delta (t - t'). \qquad (47)$$

In the limit that $T \to 0$, $\sigma_{\alpha\beta}^f$ vanishes, and the Navier–Stokes equations are recovered.

We begin our analysis with a general description of the dynamics of the LB equation, based on a Chapman–Enskog expansion (Sect. 3.2). Then we consider the equilibrium distribution for the D3Q19 model (Sect. 3.3), followed by deterministic (Sect. 3.4) and stochastic (Sect. 3.5) collision operators. Finally, we consider the connection of the fluctuating LB model to statistical mechanics (Sect. 3.6) and the effects of external forces (Sect. 3.7).

3.2 Chapman–Enskog Expansion

The Navier–Stokes description of a fluid is more coarse-grained than the original LB equation, and to connect the microscopic scales with the hydrodynamic scales we follow a standard asymptotic analysis [103]. We first introduce a dimensionless scaling parameter $\varepsilon \ll 1$ and write

$$\mathbf{r}_1 = \varepsilon \mathbf{r}. \tag{48}$$

The idea is to measure spatial positions with a ruler that has such a coarse scale that details at the lattice level are not resolved. The position $\mathbf{r}_1$ then corresponds to the number read off from this coarse-grained ruler; for example instead of talking about 1,000 nm, we talk about 1 μm. For two points to be distant on the hydrodynamic scale, it is not sufficient that $|\Delta \mathbf{r}|$ is large, but rather that $|\Delta \mathbf{r}_1|$ is large. However, from the perspective of practical computation, the degree of coarse graining is never as extensive as implied by our analysis; the calculations would take far too long. Instead there is usually only a few grid points separating the lattice scale from the smallest hydrodynamic scale. Surprisingly the LB method can be quite accurate, even in these circumstances [99, 104].

In a similar way, we can also introduce a coarse-grained clock for the time variable, and write

$$t_1 = \varepsilon t. \tag{49}$$

The fact that we choose the same factor ε for both space and time is related to the typical scaling of wave-like phenomena, where the time scale of a process is *linearly* proportional to the corresponding length scale. However, hydrodynamics also includes diffusion of momentum, where the time scale is proportional to the *square* of the length scale. These processes occur on a much longer time scale, and to capture the slow dynamics we introduce a second clock that is even more coarse-grained,

$$t_2 = \varepsilon^2 t. \tag{50}$$

We can therefore distinguish between "short times" on the hydrodynamic scale, characterized by $t_s = t_1/\varepsilon$, and "long times," where $t_l = t_2/\varepsilon^2$. Both t_s and t_l are implicitly large on the lattice scale, with the hydrodynamic limit being reached as $\varepsilon \to 0$. But once again, practical computation limits the separation between the time scales h, t_s, and t_l to one or two orders of magnitude each.

In the "multi-time scale" analysis, the LB population densities may be considered to be functions of the coarse-grained position and times, $\mathbf{r}_1$, t_1, and t_2; $n_i \equiv n_i(\mathbf{r}_1, t_1, t_2)$. When the algorithm proceeds by one time step, $t \to t + h$, $t_1 \to t_1 + \varepsilon h$, and $t_2 \to t_2 + \varepsilon^2 h$. The LB equation in terms of the coarse-grained variables is then,

$$n_i(\mathbf{r}_1 + \varepsilon \mathbf{c}_i h, t_1 + \varepsilon h, t_2 + \varepsilon^2 h) - n_i(\mathbf{r}_1, t_1, t_2) = \Delta_i(\mathbf{n}(\mathbf{r}_1, t_1, t_2)). \tag{51}$$

The population densities are slowly varying functions of coarse-grained variables, and we may obtain hydrodynamic behavior by a Taylor expansion of n_i (51) to second order in powers of ε:

$$n_i(\mathbf{x} + \delta \mathbf{x}) = n_i(\mathbf{x}) + \sum_k \frac{\partial n_i}{\partial x_k} \delta x_k + \frac{1}{2} \sum_{kl} \frac{\partial^2 n_i}{\partial x_k \partial x_l} \delta x_k \delta x_l + \cdots, \tag{52}$$

where we use $\mathbf{x}$ to indicate the coarse-grained variables, $[\mathbf{r}_1, t_1, t_2]$. Since the distribution function itself depends on the degree of coarse-graining, we must take the ε

dependence of the n_i and Δ_i into account as well:

$$n_i = n_i^{(0)} + \varepsilon n_i^{(1)} + O(\varepsilon^2), \tag{53}$$

$$\Delta_i = \Delta_i^{(0)} + \varepsilon \Delta_i^{(1)} + \varepsilon^2 \Delta_i^{(2)} + O(\varepsilon^3). \tag{54}$$

The conservation laws for mass and momentum must hold independently of the value of ε, and thus at every order k:

$$\sum_i \Delta_i^{(k)} = \sum_i \Delta_i^{(k)} \mathbf{c}_i = 0. \tag{55}$$

Inserting these expansions into (51), and collecting terms at different orders of ε, we obtain:

- At order ε^0,

$$\Delta_i^{(0)} = 0 \tag{56}$$

- At order ε^1,

$$(\partial_{t_1} + \mathbf{c}_i \cdot \partial_{\mathbf{r}_1}) n_i^{(0)} = h^{-1} \Delta_i^{(1)} \tag{57}$$

- At order ε^2,

$$\partial_{t_2} n_i^{(0)} + \frac{h}{2} (\partial_{t_1} + \mathbf{c}_i \cdot \partial_{\mathbf{r}_1})^2 n_i^{(0)} + (\partial_{t_1} + \mathbf{c}_i \cdot \partial_{\mathbf{r}_1}) n_i^{(1)} = h^{-1} \Delta_i^{(2)} \tag{58}$$

Subsequently, it will prove useful to eliminate the second occurrence of $n_i^{(0)}$ from (58), by using (57):

$$\partial_{t_2} n_i^{(0)} + \frac{1}{2} (\partial_{t_1} + \mathbf{c}_i \cdot \partial_{\mathbf{r}_1}) \left(n_i^{\star(1)} + n_i^{(1)} \right) = h^{-1} \Delta_i^{(2)}, \tag{59}$$

where $n_i^\star = n_i + \Delta_i$ is the post-collision population in direction i.

The multi-time-scale expansion of (51) is based on the physical time-scale separation between collisions ($t \sim h$), sound propagation ($t \sim h/\varepsilon$), and momentum diffusion ($t \sim h/\varepsilon^2$). Equations (56)–(58) make the implicit assumption that these three relaxations can be considered separately, which allows the collision operator at order $k+1$ to be calculated from the distribution functions at order k. In essence, the collision dynamics at order $k+1$ is slaved to the lower-order distributions. The zeroth-order collision operator must be a function of $\mathbf{n}^{(0)}$ only,

$$\Delta_i^{(0)} = \Delta_i(\mathbf{n}^{(0)}), \tag{60}$$

which, in conjunction with (56), shows that $\mathbf{n}^{(0)}$ is a collisional invariant; thus we can associate $\mathbf{n}^{(0)}$ with the equilibrium distribution $\mathbf{n}^{eq}$ [105]. In order to avoid spurious conserved quantities, the equilibrium distribution should be a function of local values of the conserved variables, ρ and $\mathbf{j}$, only. In a homogeneous system, with fixed mass and momentum densities, $\mathbf{n}^{eq}(\rho,\mathbf{j}) = \mathbf{n}^{(0)}(\rho,\mathbf{j})$ is stationary in time. A stochastic collision operator (see Sect. 3.5) cannot satisfy (56) and therefore must enter the Chapman–Enskog expansion at order ε.

From (53) we can derive analogous ε expansions for ρ and $\mathbf{j}$,

$$\rho = \rho^{(0)} + \varepsilon\rho^{(1)} + \varepsilon^2\rho^{(2)} + O(\varepsilon^3), \tag{61}$$

$$\mathbf{j} = \mathbf{j}^{(0)} + \varepsilon\mathbf{j}^{(1)} + \varepsilon^2\mathbf{j}^{(2)} + O(\varepsilon^3). \tag{62}$$

However, inserting these expansions into $n_i^{(0)}(\rho,\mathbf{j})$, shows that

$$0 = \rho^{(1)} = \rho^{(2)} = \cdots, \tag{63}$$

$$0 = \mathbf{j}^{(1)} = \mathbf{j}^{(2)} = \cdots; \tag{64}$$

otherwise $n_i^{(0)}$ would have contributions of order ε and above, in contradiction to (56). The mass and momentum densities can therefore be defined as moments of the equilibrium distribution as well,

$$\sum_i n_i^{eq} = \rho, \tag{65}$$

$$\sum_i n_i^{eq}\mathbf{c}_i = \mathbf{j}. \tag{66}$$

We can analyze the dynamics of the LB model on large length and time scales by taking moments of (57) and (59) with respect to the LB velocity set $\mathbf{c}_i$. From the zeroth moment, $\sum_i \cdots$, we obtain the continuity equation on the t_1 time scale (55),

$$\partial_{t_1}\rho + \partial_{1\alpha}j_\alpha = 0, \tag{67}$$

and incompressibility on the t_2 time scale (55), (63), and (64)

$$\partial_{t_2}\rho = 0. \tag{68}$$

In (67) we have used the shorthand notation $\partial_{1\alpha}$ for the α component of the spatial derivative $\partial_{\mathbf{r}_1}$.

The first moment, $\sum_i \cdots c_{i\alpha}$, leads to momentum conservation equations on both time scales (55), (63), and (64):

$$\partial_{t_1}j_\alpha + \partial_{1\beta}\pi_{\alpha\beta}^{(0)} = 0, \tag{69}$$

$$\partial_{t_2}j_\alpha + \frac{1}{2}\partial_{1\beta}\left(\pi_{\alpha\beta}^{\star(1)} + \pi_{\alpha\beta}^{(1)}\right) = 0, \tag{70}$$

where $\pi_{\alpha\beta}$ is the momentum flux or second moment,[1]

$$\pi_{\alpha\beta} = \sum_i n_i c_{i\alpha} c_{i\beta}. \tag{71}$$

Momentum is conserved on both the t_1 and t_2 time scales, because, in the hydro-dynamic limit, the coupling between acoustic and diffusive modes is very weak. First, sound waves propagate with negligible viscous damping; then the residual pressure field in a nearly incompressible fluid relaxes by momentum diffusion. We can write the conservation laws on each time scale separately, as in (69) and (70), or combine them into a single equation in the lattice-scale variables $\mathbf{r} = \mathbf{r}_1/\varepsilon$ and $t = t_1/\varepsilon = t_2/\varepsilon^2$. The hydrodynamic fields depend on $\mathbf{r}$ and t parametrically, through their dependence on the coarse-grained variables $\mathbf{r}_1, t_1, t_2$. Using ∂_α for a component of $\partial_\mathbf{r}$, we have

$$\partial_\alpha = \varepsilon\partial_{1\alpha}, \tag{72}$$
$$\partial_t = \varepsilon\partial_{t_1} + \varepsilon^2\partial_{t_2}. \tag{73}$$

The combined equations for the mass and momentum densities on the lattice space and time scales are then:

$$\partial_t \rho + \partial_\alpha j_\alpha = 0, \tag{74}$$
$$\partial_t j_\alpha + \partial_\beta \pi_{\alpha\beta}^{eq} + \frac{1}{2}\partial_\beta\left(\pi_{\alpha\beta}^{\star neq} + \pi_{\alpha\beta}^{neq}\right) = 0, \tag{75}$$

where from (53), $\pi_{\alpha\beta}^{eq} = \pi_{\alpha\beta}^{(0)}$ and $\pi_{\alpha\beta}^{neq} = \varepsilon\pi_{\alpha\beta}^{(1)}$.

Finally, we can derive a relation between the pre-collision and post-collision momentum fluxes, $\pi_{\alpha\beta}$ and $\pi_{\alpha\beta}^\star$, by taking the second moment of (57):

$$\partial_{t_1}\pi_{\alpha\beta}^{(0)} + \partial_{1\gamma}\Phi_{\alpha\beta\gamma}^{(0)} = h^{-1}\left(\pi_{\alpha\beta}^{\star(1)} - \pi_{\alpha\beta}^{(1)}\right), \tag{76}$$

where $\Phi_{\alpha\beta\gamma}$ is the third moment of the distribution,

$$\Phi_{\alpha\beta\gamma} = \sum_i n_i c_{i\alpha} c_{i\beta} c_{i\gamma}. \tag{77}$$

We note that $\pi_{\alpha\beta}^{eq}$ is a collisional invariant and therefore remains unchanged by the collision process. In terms of the lattice variables,

$$\pi_{\alpha\beta}^{\star neq} = \pi_{\alpha\beta}^{neq} + h\left(\partial_t\pi_{\alpha\beta}^{eq} + \partial_\gamma\Phi_{\alpha\beta\gamma}^{eq}\right). \tag{78}$$

---

[1] There is a notational inconsistency in [102]. In (71) and (73) of that paper the superscript "neq" should be replaced by a superscript 1, and in (79) $Q_{\alpha\beta}$ should be $Q_{\alpha\beta}^1$.

Equation (74) shows that continuity (43) is automatically satisfied by any LB model. The Navier–Stokes equation (44) will be satisfied, *if* we succeed in ensuring that the Euler stress $\rho c_s^2 \delta_{\alpha\beta} + \rho u_\alpha u_\beta$, the Newtonian viscous stress, $\sigma_{\alpha\beta}$ (45), and the fluctuating stress $\sigma_{\alpha\beta}^f$ (47) are given correctly by the sum of the momentum fluxes in (75). Since $\pi_{\alpha\beta}^{eq}$ depends *only* on ρ and $\mathbf{j}$, it must be identified with the Euler stress:

$$\pi_{\alpha\beta}^{eq} = \rho c_s^2 \delta_{\alpha\beta} + \rho u_\alpha u_\beta. \tag{79}$$

The viscous stress and fluctuating stresses must then be contained in $(\pi_{\alpha\beta}^{\star neq} + \pi_{\alpha\beta}^{neq})/2$.

This is about as far as we can go in complete generality. In order to proceed further we need to consider specific equilibrium distributions and collision operators. The results of this subsection suggest the following approach towards constructing an LB method which (asymptotically) simulates the fluctuating Navier–Stokes equations:

- Find a set of equilibrium populations n_i^{eq} such that:

  –

$$\sum_i n_i^{eq} = \rho . \tag{80}$$

  –

$$\sum_i n_i^{eq} \mathbf{c}_i = \mathbf{j} . \tag{81}$$

  –

$$\sum_i n_i^{eq} c_{i\alpha} c_{i\beta} = \rho c_s^2 \delta_{\alpha\beta} + \rho u_\alpha u_\beta . \tag{82}$$

- Find a collision operator Δ_i with the properties:

  –

$$\sum_i \Delta_i = 0 . \tag{83}$$

  –

$$\sum_i \Delta_i \mathbf{c}_i = 0 . \tag{84}$$

  – The nonequilibrium momentum flux $(\pi_{\alpha\beta}^{\star neq} + \pi_{\alpha\beta}^{neq})/2$ must be connected with the sum of viscous and fluctuating stresses.

In the following subsections, we will follow this procedure for the three-dimensional D3Q19 model.

3.3 D3Q19 Model I: Equilibrium Populations

Early LB models [78–80] inherited their equilibrium distributions from LGCA, along with macroscopic manifestations of the broken Galilean invariance: an incorrect advection velocity and a velocity-dependent pressure. Subsequently a new equilibrium distribution was proposed that restored Galilean invariance at the macroscopic level [97, 106], but with the loss of the connection to statistical mechanics. The idea was to ensure that the first few moments of n_i^{eq} matched those derived from the Maxwell–Boltzmann distribution for a dilute gas [105],

$$n(\mathbf{c}|\rho, \mathbf{u}, T) = \rho \left(\frac{m_p}{2\pi k_B T}\right)^{3/2} \exp\left[-\frac{m_p}{2k_B T} (\mathbf{c} - \mathbf{u})^2\right] : \qquad (85)$$

specifically;

$$\int d^3\mathbf{c}\, n(\mathbf{c}) = \rho, \qquad (86)$$

$$\int d^3\mathbf{c}\, n(\mathbf{c}) c_\alpha = \rho u_\alpha, \qquad (87)$$

$$\int d^3\mathbf{c}\, n(\mathbf{c}) c_\alpha c_\beta = \frac{\rho k_B T}{m_p} \delta_{\alpha\beta} + \rho u_\alpha u_\beta = \rho c_s^2 \delta_{\alpha\beta} + \rho u_\alpha u_\beta. \qquad (88)$$

With these moments the Euler hydrodynamic equations [cf. (43) and (44)],

$$\partial_t \rho + \partial_\alpha j_\alpha = 0, \qquad (89)$$
$$\partial_t j_\alpha + \partial_\beta \left(\rho c_s^2 \delta_{\alpha\beta} + \rho u_\alpha u_\beta\right) = 0, \qquad (90)$$

may be derived from the continuum version of the Chapman–Enskog expansion [105]. The viscous stress arises from the non-equilibrium distribution (cf. Sect. 3.2).

An expansion of the Maxwell–Boltzmann equilibrium distribution (85) at low velocities suggests the following ansatz [97, 106] for the discrete velocity equilibrium,

$$n_i^{eq}(\rho, \mathbf{u}) = a^{c_i} \rho \left(1 + A\mathbf{u} \cdot \mathbf{c}_i + B(\mathbf{u} \cdot \mathbf{c}_i)^2 + Cu^2\right) \qquad (91)$$

with suitably adjusted coefficients a^{c_i}, A, B, and C. The rationale for (91) is that the equilibrium momentum flux is quadratic in the flow velocity $\mathbf{u}$ (88); it therefore makes sense to construct a similar form for n_i^{eq}. A drawback of (91) is that n_i^{eq} may be negative if u becomes sufficiently large. This can be avoided by more general equilibrium distributions, which are equivalent to (91) up to order u^2 [107–109].

The prefactors $a^{c_i} > 0$ are normalized such that

$$\sum_i a^{c_i} = 1, \qquad (92)$$

which ensures that (65) is satisfied in the special case $\mathbf{u} = 0$. The notation in (91) was chosen in order to indicate explicitly that the weights depend only on the absolute value of the speed c_i, but not its direction; this follows from the rotational symmetries of the LB model. The coefficients A, B, C are here independent of c_i. There are other LB models, like D3Q18 [106], where this condition is not imposed, and A, B, and C depend on c_i as well; however, such models are only hydrodynamically correct in the incompressible limit [98], and cannot be straightforwardly interpreted in terms of statistical mechanics (see Sect. 3.6). We will not consider such models.

In a cubic lattice, symmetry dictates the following relations for the low-order velocity moments of the weights,

$$\sum_i a^{c_i} c_{i\alpha} = 0, \tag{93}$$

$$\sum_i a^{c_i} c_{i\alpha} c_{i\beta} = C_2 \delta_{\alpha\beta}, \tag{94}$$

$$\sum_i a^{c_i} c_{i\alpha} c_{i\beta} c_{i\gamma} = 0, \tag{95}$$

$$\sum_i a^{c_i} c_{i\alpha} c_{i\beta} c_{i\gamma} c_{i\delta} = C_4' \delta_{\alpha\beta\gamma\delta} + C_4 \left(\delta_{\alpha\beta} \delta_{\gamma\delta} + \delta_{\alpha\gamma} \delta_{\beta\delta} + \delta_{\alpha\delta} \delta_{\beta\gamma} \right), \tag{96}$$

where the values of the parameters C_2, C_4 and C_4' depend on the details of the choice of the coefficients a^{c_i}. The tensor $\delta_{\alpha\beta\gamma\delta}$ is unity when $\alpha = \beta = \gamma = \delta$ and zero otherwise. Rotational invariance of the stress tensor requires that $C_4' = 0$.

The results in (92)–(96) allow us to calculate the moments of (91) up to second order. Consistency with the mass density, momentum density and Euler stress for a given ρ and $\mathbf{u}$, uniquely determines the equilibrium distribution,

$$n_i^{\text{eq}} (\rho, \mathbf{u}) = a^{c_i} \rho \left(1 + \frac{\mathbf{u} \cdot \mathbf{c}_i}{c_s^2} + \frac{(\mathbf{u} \cdot \mathbf{c}_i)^2}{2 c_s^4} - \frac{u^2}{2 c_s^2} \right), \tag{97}$$

with the speed of sound $c_s^2 = C_2$, and the weights adjusted such that $C_4' = 0$ and $C_4 = C_2^2$. These two latter conditions, together with the normalization condition (92), form a set of three equations for the coefficients a^{c_i}. Therefore at least three speeds, or three shells of neighbors, are needed to satisfy the constraints. We consider the D3Q19 model, which incorporates the three smallest speeds on a simple-cubic lattice. Here one obtains $a^0 = 1/3$ for the stationary particles, $a^1 = 1/18$ for the six nearest-neighbor directions, and $a^{\sqrt{2}} = 1/36$ for the 12 next-nearest neighbors. The speed of sound is then $c_s^2 = (1/3)(b/h)^2$.

We now turn back to the results of the previous subsection, since the explicit form of n_i^{eq} allows us to pursue the analysis further. We first calculate the equilibrium third-order moment (77) using (97):

$$\Phi_{\alpha\beta\gamma}^{\text{eq}} = \rho c_s^2 \left(u_\alpha \delta_{\beta\gamma} + u_\beta \delta_{\alpha\gamma} + u_\gamma \delta_{\alpha\beta} \right). \tag{98}$$

In fact (98) is model independent to order u^2, since only the linear term in $\mathbf{u} \cdot \mathbf{c}_i$ contributes. To close the hydrodynamic equations for the mass and momentum densities [(74) and (75)] we need expressions for the pre-collision and post-collision momentum fluxes, $\pi_{\alpha\beta}^{\star\text{neq}}$ and $\pi_{\alpha\beta}^{\text{neq}}$. From (76) we can obtain an expression for $\pi_{\alpha\beta}^{\star\text{neq}} - \pi_{\alpha\beta}^{\text{neq}}$ in terms of the velocity gradient,

$$\rho c_s^2 \left(\partial_{1\alpha} u_\beta + \partial_{1\beta} u_\alpha \right) = h^{-1} \left(\pi_{\alpha\beta}^{\star(1)} - \pi_{\alpha\beta}^{(1)} \right), \tag{99}$$

where we have used (67) and (69) to rewrite the time derivative of $\pi_{\alpha\beta}^{(0)}$ in terms of spatial derivatives of ρ and $\mathbf{u}$. In arriving at (99), we have neglected terms of order u^3, consistent with the low Mach number limit we are considering. Finally, (99) can be rewritten in terms of the unscaled variables,

$$\pi_{\alpha\beta}^{\star\text{neq}} - \pi_{\alpha\beta}^{\text{neq}} = h\rho c_s^2 \left(\partial_\alpha u_\beta + \partial_\beta u_\alpha \right). \tag{100}$$

To obtain a further relation for the non-equilibrium momentum fluxes, we must consider the collision operator in more detail.

3.4 D3Q19 Model II: Deterministic Collision Operator

In a deterministic model, the collision operator Δ_i is a unique function of the distribution $\mathbf{n}$. Therefore, we can obtain the Chapman–Enskog ordering of Δ_i via a Taylor expansion with respect to $\mathbf{n}$:

$$\Delta_i (\mathbf{n}) = \Delta_i \left(\mathbf{n}^{(0)} + \varepsilon \mathbf{n}^{(1)} + \varepsilon^2 \mathbf{n}^{(2)} + \cdots \right)$$

$$= \Delta_i \left(\mathbf{n}^{(0)} \right) + \varepsilon \sum_j \left(\frac{\partial \Delta_i}{\partial n_j} \right) \Bigg|_{\mathbf{n}^{(0)}} n_j^{(1)} + O \left(\varepsilon^2 \right). \tag{101}$$

The analysis of Sect. 3.2 has shown that $\Delta_i(\mathbf{n}^{(0)}) = 0$ (60), and that hydrodynamic behavior is determined by the order ε^1 collision operator,

$$\Delta_i^{(1)} = \sum_j \left(\frac{\partial \Delta_i}{\partial n_j} \right) \Bigg|_{\mathbf{n}^{(0)}} n_j^{(1)}. \tag{102}$$

Although $\Delta_i^{(2)}$ appears at second order in the Chapman–Enskog expansion (58), it makes no contribution to the change in mass and momentum densities (55). However, $\Delta_i^{(1)}$ contributes to a first-order change in the viscous stress (78), which enters into the momentum equation at second order (70). It is therefore reasonable to construct the collision operator with the form of $\Delta_i^{(1)}$:

$$\Delta_i = \sum_j \mathcal{L}_{ij} n_j^{\text{neq}}, \tag{103}$$

where $\mathcal{L}_{ij}$ is a matrix of constant coefficients. Thus to lowest order in ε, the collision process is a linear transformation between the non-equilibrium distributions for each velocity:

$$n_i^{\star\text{neq}} = \sum_j \left(\delta_{ij} + \mathcal{L}_{ij} \right) n_j^{\text{neq}}. \tag{104}$$

The simplest such collision operator is the lattice BGK (Bhatnagar–Gross–Krook) model [77], $\mathcal{L}_{ij} = -\delta_{ij}/\tau$, where the collisional relaxation time τ is related to the viscosity. Here we will work within the more general framework of the multi-relaxation time (MRT) model [110], for which the lattice BGK model is a special case.

Polynomials in the dimensionless velocity vectors, $\hat{\mathbf{c}}_i = \mathbf{c}_i/c$ ($c = b/h$), form a basis for a diagonal representation of $\mathcal{L}_{ij}$ [110], which allows for a more general and stable LB model with the same level of computational complexity as the BGK version [111]. Orthogonal basis vectors, $\mathbf{e}_k$, are constructed from outer products of the vectors $\hat{\mathbf{c}}_i$. For example:

$$e_{0i} = 1, \tag{105}$$

$$e_{1i} = \hat{c}_{ix}, \tag{106}$$

$$e_{2i} = \hat{c}_{iy}, \tag{107}$$

$$e_{3i} = \hat{c}_{iz}. \tag{108}$$

There are six quadratic polynomials, which are given in Table 1 as basis vectors $\mathbf{e}_4 - \mathbf{e}_9$. A Gram–Schmidt procedure ensures that all the basis vectors are mutually

Table 1 Basis vectors of the D3Q19 model. Each row corresponds to a different basis vector, with the actual polynomial in $\hat{c}_{i\alpha} = c_{i\alpha}/c$ shown in the second column. The normalizing factor for each basis vector is in the third column. The polynomials form an orthogonal set when $q^{ci} = a^{ci}$ (109)

k	e_{ki}	w_k
0	1	1
1	$\hat{c}_{ix}$	1/3
2	$\hat{c}_{iy}$	1/3
3	$\hat{c}_{iz}$	1/3
4	$\hat{c}_i^2 - 1$	2/3
5	$3\hat{c}_{ix}^2 - \hat{c}_i^2$	4/3
6	$\hat{c}_{iy}^2 - \hat{c}_{iz}^2$	4/9
7	$\hat{c}_{ix}\hat{c}_{iy}$	1/9
8	$\hat{c}_{iy}\hat{c}_{iz}$	1/9
9	$\hat{c}_{iz}\hat{c}_{ix}$	1/9
10	$(3\hat{c}_i^2 - 5)\hat{c}_{ix}$	2/3
11	$(3\hat{c}_i^2 - 5)\hat{c}_{iy}$	2/3
12	$(3\hat{c}_i^2 - 5)\hat{c}_{iz}$	2/3
13	$(\hat{c}_{iy}^2 - \hat{c}_{iz}^2)\hat{c}_{ix}$	2/9
14	$(\hat{c}_{iz}^2 - \hat{c}_{ix}^2)\hat{c}_{iy}$	2/9
15	$(\hat{c}_{ix}^2 - \hat{c}_{iy}^2)\hat{c}_{iz}$	2/9
16	$3\hat{c}_i^4 - 6\hat{c}_i^2 + 1$	2
17	$(2\hat{c}_i^2 - 3)(3\hat{c}_{ix}^2 - \hat{c}_i^2)$	4/3
18	$(2\hat{c}_i^2 - 3)(\hat{c}_{iy}^2 - \hat{c}_{iz}^2)$	4/9

orthogonal with respect to a set of positive weights, $q^{c_i} > 0$,

$$\sum_i q^{c_i} e_{ki} e_{li} = w_k \delta_{kl}. \tag{109}$$

The weights are restricted by the same symmetries as the coefficients in the equilibrium distribution a^{c_i}, but are not necessarily the same; in the D3Q19 model there are then three independent values of q^{c_i}. The normalization factors, $w_k > 0$, are related to the choice of basis vectors

$$w_k = \sum_i q^{c_i} e_{ki}^2. \tag{110}$$

Within the D3Q19 model, polynomials up to second order are complete, but at third order there is some deflation; for example, $\hat{c}_{ix}^3$ is equivalent to $\hat{c}_{ix}$. In fact, there are only six independent third-order and three independent fourth-order polynomials in the D3Q19 model. Beyond fourth order, all polynomials deflate to lower orders, so the basis vectors in Table 1 form a complete set for the D3Q19 model.

The basis vectors can be used to construct a complete set of moments of the LB distribution,

$$m_k = \sum_i e_{ki} n_i, \tag{111}$$

which allows for a diagonal representation of the collision operator [110, 112], as will be made clear later. Hydrodynamic variables are related to the moments up to quadratic order in $\hat{c}_i$ (cf. Table 1):

$$\rho = m_0, \tag{112}$$

$$j_x = m_1 c, \tag{113}$$

$$j_y = m_2 c, \tag{114}$$

$$j_z = m_3 c, \tag{115}$$

$$\pi_{xx} = (m_0 + m_4 + m_5)c^2/3, \tag{116}$$

$$\pi_{yy} = (2m_0 + 2m_4 - m_5 + 3m_6)c^2/6, \tag{117}$$

$$\pi_{zz} = (2m_0 + 2m_4 - m_5 - 3m_6)c^2/6, \tag{118}$$

$$\pi_{xy} = m_7 c^2, \tag{119}$$

$$\pi_{yz} = m_8 c^2, \tag{120}$$

$$\pi_{zx} = m_9 c^2. \tag{121}$$

There are additional degrees of freedom in the D3Q19 model beyond those required for the conserved variables and stresses (112)–(121). These "kinetic" or "ghost" [101] moments do not play a role in the large-scale dynamics [102], but they are important for proper thermalization [101] and near boundaries [113].

The basis vectors in Table 1 are complete but not unique. Besides trivial variations in the Gram–Schmidt orthogonalization, there is a substantive difference that depends on the choice of the weighting factors q^{c_i}: these factors determine both the result of the orthogonalization procedure, as well as the back transformation from moments m_k to populations n_i. This is most easily seen from the observation that (109) can be rewritten as the standard orthonormality relation [102]

$$\sum_i \hat{e}_{ki}\hat{e}_{li} = \delta_{kl}, \tag{122}$$

where we have introduced the orthonormal basis vectors

$$\hat{e}_{ki} = \sqrt{\frac{q^{c_i}}{w_k}}\, e_{ki}. \tag{123}$$

Equation (122) implies the backward relation

$$\sum_k \hat{e}_{ki}\hat{e}_{kj} = \delta_{ij}, \tag{124}$$

or, in terms of unnormalized basis vectors,

$$\sum_k w_k^{-1} e_{ki} e_{kj} = \frac{1}{q^{c_i}}\delta_{ij}. \tag{125}$$

In the normalized basis the transformations between distribution and moments are

$$\hat{m}_k = \frac{m_k}{\sqrt{w_k}} = \sum_i \hat{e}_{ki}\frac{n_i}{\sqrt{q^{c_i}}} = \sum_i \hat{e}_{ki}\hat{n}_i, \tag{126}$$

$$\hat{n}_i = \frac{n_i}{\sqrt{q^{c_i}}} = \sum_k \hat{e}_{ki}\frac{m_k}{\sqrt{w_k}} = \sum_k \hat{e}_{ki}\hat{m}_k; \tag{127}$$

we will make use of these relations in Sect. 3.5. The analog of (104) for the normalized basis is

$$\hat{n}_i^{\star\text{neq}} = \sum_j \left(\delta_{ij} + \hat{\mathcal{L}}_{ij}\right)\hat{n}_j^{\text{neq}}, \tag{128}$$

with

$$\hat{\mathcal{L}}_{ij} = \sqrt{\frac{q^{c_j}}{q^{c_i}}}\,\mathcal{L}_{ij}. \tag{129}$$

In terms of unnormalized basis vectors the back transformation is given by

$$n_i = q^{c_i}\sum_k w_k^{-1} e_{ki} m_k. \tag{130}$$

The most obvious choice is to set $q^{c_i} = 1$ [110, 112], but then the basis vectors of the kinetic modes, $\mathbf{e}_{10}$–$\mathbf{e}_{18}$, are not orthogonal to the equilibrium distribution, and the moments m_{10}–m_{18} have both equilibrium and non-equilibrium contributions [110, 112]. The statistical mechanical connection is more straightforward if the weights q^{c_i} are matched to the weights in the equilibrium distribution, setting $q^{c_i} = a^{c_i}$; this eliminates the projection of n_i^{eq} on the kinetic moments. The weighted orthogonality relation defines a different but equivalent set of basis vectors to those given in [110, 112], and these are the ones given in Table 1. A comparison of the two sets of basis vectors can be found in [114].

The basis vectors can be used to construct a collision operator that automatically satisfies all the lattice symmetries,

$$\hat{L}_{ij} = \sum_k \lambda_k \hat{e}_{ki} \hat{e}_{kj}, \tag{131}$$

which is a symmetric matrix, while $\mathcal{L}_{ij}$, in general, is not symmetric. The orthogonality of the basis vectors ensures that each moment relaxes independently under the action of the linearized collision operator,

$$\hat{m}_k^{\star \text{neq}} = \gamma_k \hat{m}_k^{\text{neq}}, \tag{132}$$

where $\gamma_k = 1 + \lambda_k$. For the conserved modes $k = 0, \ldots, 3$ the value γ_k is immaterial, since $m_k^{\star \text{neq}} = m_k^{\text{neq}} = 0$. For the other modes, $k > 3$, linear stability requires that

$$|\gamma_k| \leq 1; \tag{133}$$

i.e., the effect of collisions must be to cause the nonequilibrium distribution to decrease rather than increase. The eigenvalues, γ_k, may be positive or negative, with $\gamma_k < 0$ corresponding to "over-relaxation."

The number of independent eigenvalues is limited by symmetry. There are at most six independent γ_k's in the D3Q19 model, corresponding to a bulk viscous mode with eigenvalue γ_v, five symmetry-related shear modes, which must have the same eigenvalue, γ_s, and nine kinetic modes, broken down into symmetry-related groups: $e_{10}-e_{12}$, $e_{13}-e_{15}$, e_{16}, $e_{17}-e_{18}$. The eigenvalues γ_s and γ_v can be related to the shear and bulk viscosities by decomposing the stress tensor into traceless-symmetric (shear) and trace (bulk) components,

$$\pi_{\alpha\beta} = \overline{\pi}_{\alpha\beta} + \frac{1}{3} \pi_{\gamma\gamma} \delta_{\alpha\beta}; \tag{134}$$

the overbar is used to denote a traceless tensor. Equation (132) implies the following relations between pre- and post-collisional stresses:

$$\overline{\pi}_{\alpha\beta}^{\star \text{neq}} = \gamma_s \overline{\pi}_{\alpha\beta}^{\text{neq}}, \tag{135}$$

$$\pi_{\alpha\alpha}^{\star \text{neq}} = \gamma_v \pi_{\alpha\alpha}^{\text{neq}}. \tag{136}$$

Additional relations between the pre- and post-collision stresses have already been provided (100):

$$\overline{\pi}_{\alpha\beta}^{\star \text{neq}} - \overline{\pi}_{\alpha\beta}^{\text{neq}} = h\rho c_s^2 \left(\overline{\partial_\alpha u_\beta} + \overline{\partial_\beta u_\alpha} \right), \tag{137}$$

$$\pi_{\alpha\alpha}^{\star neq} - \pi_{\alpha\alpha}^{neq} = 2h\rho c_s^2 \partial_\alpha u_\alpha. \tag{138}$$

Equations (135)–(138) can be solved to relate the pre- and post-collision stresses to the velocity gradient:

$$\overline{\pi}_{\alpha\beta}^{neq} = -\frac{h\rho c_s^2}{1 - \gamma_s} \left(\overline{\partial_\alpha u_\beta} + \overline{\partial_\beta u_\alpha} \right), \tag{139}$$

$$\overline{\pi}_{\alpha\beta}^{\star neq} = -\frac{h\rho c_s^2 \gamma_s}{1 - \gamma_s} \left(\overline{\partial_\alpha u_\beta} + \overline{\partial_\beta u_\alpha} \right), \tag{140}$$

$$\pi_{\alpha\alpha}^{neq} = -\frac{2h\rho c_s^2}{1 - \gamma_v} \partial_\alpha u_\alpha, \tag{141}$$

$$\pi_{\alpha\alpha}^{\star neq} = -\frac{2h\rho c_s^2 \gamma_v}{1 - \gamma_v} \partial_\alpha u_\alpha. \tag{142}$$

From (75) we then find the usual Newtonian form for the viscous stress, and can identify the shear and bulk viscosities:

$$\eta = \frac{h\rho c_s^2}{2} \frac{1 + \gamma_s}{1 - \gamma_s}, \tag{143}$$

$$\eta_v = \frac{h\rho c_s^2}{3} \frac{1 + \gamma_v}{1 - \gamma_v}. \tag{144}$$

Lattice symmetry dictates that there are at most four independent eigenvalues of the kinetic modes: (see Table 1): γ_{3a} (modes 10–12), γ_{3b} (modes 13–15), γ_{4a} (mode 16), and γ_{4b} (modes 17–18). In a number of implementations of the MRT model [99, 100, 106, 115] the kinetic eigenvalues are set to zero, so that these modes are projected out by the collision operator, although they reoccur at the next time step. Recently, it has been shown that the kinetic eigenvalues can be tuned to improve the accuracy of the boundary conditions at solid surfaces [113]. A useful simplification is to use only two independent relaxation rates, with $\gamma_v = \gamma_s = \gamma_{4a} = \gamma_{4b} = \gamma_e$ and $\gamma_{3a} = \gamma_{3b} = \gamma_o$. The optimal boundary conditions are obtained with specific relations between γ_e and γ_o [113, 114].

3.5 D3Q19 Model III: Thermal Noise

In the fluctuating LB model [98, 100], thermal noise is included by adding a stochastic contribution, Δ_i', to the collision operator:

$$\Delta_i - \sum_j \mathcal{L}_{ij} n_j^{neq} + \Delta_i'. \tag{145}$$

The collision operator must still conserve mass and momentum exactly,

$$\sum_i \Delta_i' = \sum_i \Delta_i' \mathbf{c}_i = 0, \tag{146}$$

while the statistical properties of Δ_i' include a vanishing mean, $\langle \Delta_i' \rangle = 0$, and a nontrivial covariance matrix, $\langle \Delta_i' \Delta_j' \rangle$, that gives the correct fluctuations at the hydrodynamic level [see (44) and (47)]:

$$\left\langle \sigma_{\alpha\beta}^{f} \sigma_{\gamma\delta}^{f} \right\rangle = \frac{2k_B T}{b^3 h} \eta_{\alpha\beta\gamma\delta}. \tag{147}$$

The stochastic collision operator is assumed to be local in space and time, so that there are no correlations between the noise at different lattice sites or at different times. The delta functions in space and time have been replaced by b^{-3} and h^{-1}, respectively, so that the double integral of (47) with respect to $\mathbf{r}'$ and t', over a small space–time region of size $b^3 h$, matches the corresponding integral of (147).

Splitting the tensor into the trace, $\sigma_{\alpha\alpha}^{f}$, and traceless, $\bar{\sigma}_{\alpha\beta}^{f}$, parts gives the equivalent relations

$$\left\langle \bar{\sigma}_{\alpha\beta}^{f} \bar{\sigma}_{\gamma\delta}^{f} \right\rangle = \frac{2k_B T \eta}{b^3 h} \left[\delta_{\alpha\gamma}\delta_{\beta\delta} + \delta_{\alpha\delta}\delta_{\beta\gamma} - \frac{2}{3}\delta_{\alpha\beta}\delta_{\gamma\delta} \right], \tag{148}$$

$$\left\langle \sigma_{\alpha\alpha}^{f} \sigma_{\beta\beta}^{f} \right\rangle = \frac{18 k_B T \eta_v}{b^3 h}, \tag{149}$$

$$\left\langle \bar{\sigma}_{\alpha\beta}^{f} \sigma_{\gamma\gamma}^{f} \right\rangle = 0. \tag{150}$$

Although temperature does not appear directly in the D3Q19 LB model, we can determine the appropriate fluctuation level through the equation of state for an isothermal ideal gas of particles of mass m_p, $k_B T = m_p c_s^2 = \mu b^3 c_s^2$ [102]. Taking into account the results for η and η_v [(143) and (144)], we can write the desired correlations in terms of the LB variables:

$$\frac{\left\langle \bar{\sigma}_{\alpha\beta}^{f} \bar{\sigma}_{\gamma\delta}^{f} \right\rangle}{\mu \rho c_s^4} = \frac{1 + \gamma_s}{1 - \gamma_s} \left[\delta_{\alpha\gamma}\delta_{\beta\delta} + \delta_{\alpha\delta}\delta_{\beta\gamma} - \frac{2}{3}\delta_{\alpha\beta}\delta_{\gamma\delta} \right], \tag{151}$$

$$\frac{\left\langle \sigma_{\alpha\alpha}^{f} \sigma_{\beta\beta}^{f} \right\rangle}{\mu \rho c_s^4} = 6 \frac{1 + \gamma_v}{1 - \gamma_v}, \tag{152}$$

$$\left\langle \bar{\sigma}_{\alpha\beta}^{f} \sigma_{\gamma\gamma}^{f} \right\rangle = 0. \tag{153}$$

The stress fluctuations $\sigma_{\alpha\beta}^{f}$ are *different* from the random stresses $\sigma_{\alpha\beta}^{r}$ that arise in the LB algorithm itself,

$$\sigma_{\alpha\beta}^{r} = \sum_i \Delta_i' c_{i\alpha} c_{i\beta}. \tag{154}$$

The reason is that $\sigma_{\alpha\beta}^{f}$ pertains to fluctuations on the t_1 time scale, which interact with the hydrodynamic flow field, while $\sigma_{\alpha\beta}^{r}$ represents added noise on the lattice time scale, h. We use the Chapman–Enskog procedure to work backwards from the known fluctuations in $\sigma_{\alpha\beta}^{f}$ to determine the covariance matrix for $\sigma_{\alpha\beta}^{r}$ [102]. The stress update rule including random noise $\sigma_{\alpha\beta}^{r}$ is [cf. (135) and (136)]

$$\bar{\pi}_{\alpha\beta}^{\star\mathrm{neq}} = \gamma_s \bar{\pi}_{\alpha\beta}^{\mathrm{neq}} + \bar{\sigma}_{\alpha\beta}^{r}, \tag{155}$$

$$\pi_{\alpha\alpha}^{\star\mathrm{neq}} = \gamma_v \pi_{\alpha\alpha}^{\mathrm{neq}} + \sigma_{\alpha\alpha}^{r}. \tag{156}$$

Equations (137) and (138) remain valid and, together with (155) and (156), can be solved for the pre- and post-collisional stresses, $\pi_{\alpha\beta}^{\mathrm{neq}}$ and $\pi_{\alpha\beta}^{\star\mathrm{neq}}$, as before:

$$\bar{\pi}_{\alpha\beta}^{\mathrm{neq}} = -\frac{h\rho c_s^2}{1-\gamma_s}\left(\overline{\partial_\alpha u_\beta} + \overline{\partial_\beta u_\alpha}\right) + \frac{1}{1-\gamma_s}\bar{\sigma}_{\alpha\beta}^{r}, \tag{157}$$

$$\bar{\pi}_{\alpha\beta}^{\star\mathrm{neq}} = -\frac{h\rho c_s^2 \gamma_s}{1-\gamma_s}\left(\overline{\partial_\alpha u_\beta} + \overline{\partial_\beta u_\alpha}\right) + \frac{1}{1-\gamma_s}\bar{\sigma}_{\alpha\beta}^{r}, \tag{158}$$

$$\pi_{\alpha\alpha}^{\mathrm{neq}} = -\frac{2h\rho c_s^2}{1-\gamma_v}\partial_\alpha u_\alpha + \frac{1}{1-\gamma_v}\sigma_{\alpha\alpha}^{r}, \tag{159}$$

$$\pi_{\alpha\alpha}^{\star\mathrm{neq}} = -\frac{2h\rho c_s^2 \gamma_v}{1-\gamma_v}\partial_\alpha u_\alpha + \frac{1}{1-\gamma_v}\sigma_{\alpha\alpha}^{r}. \tag{160}$$

Comparing (75) with (44) we can read off the relations between the hydrodynamic fluctuations and the random noise,

$$\bar{\sigma}_{\alpha\beta}^{f} = -\frac{1}{1-\gamma_s}\bar{\sigma}_{\alpha\beta}^{r}, \tag{161}$$

$$\sigma_{\alpha\alpha}^{f} = -\frac{1}{1-\gamma_v}\sigma_{\alpha\alpha}^{r}. \tag{162}$$

Therefore, the random noise inserted at the microscopic (LB) level must have the following covariances:

$$\frac{\left\langle \bar{\sigma}_{\alpha\beta}^{r} \bar{\sigma}_{\gamma\delta}^{r} \right\rangle}{\mu\rho c_s^4} = (1-\gamma_s^2)\left[\delta_{\alpha\gamma}\delta_{\beta\delta} + \delta_{\alpha\delta}\delta_{\beta\gamma} - \frac{2}{3}\delta_{\alpha\beta}\delta_{\gamma\delta}\right], \tag{163}$$

$$\frac{\left\langle \sigma_{\alpha\alpha}^{r} \sigma_{\beta\beta}^{r} \right\rangle}{\mu\rho c_s^4} = 6(1-\gamma_v^2), \tag{164}$$

$$\left\langle \bar{\sigma}_{\alpha\beta}^{r} \sigma_{\gamma\gamma}^{r} \right\rangle = 0. \tag{165}$$

The random stress has a typical amplitude of $\sqrt{\mu\rho}c_s^2$ and is obtained from the second-order moment of the fluctuations in n_i^{neq}. Therefore, a typical fluctuation in the population density is of order $\sqrt{\mu\rho}$. Combining this scaling with (126) and (127), suggests dimensionless variables

$$\hat{n}_i = \frac{n_i}{\sqrt{a^{c_i}\mu\rho}}, \qquad (166)$$

$$\hat{m}_k = \frac{m_k}{\sqrt{w_k\mu\rho}}, \qquad (167)$$

which transform using the symmetric basis vectors defined in (123),

$$\hat{m}_k = \sum_i \hat{e}_{ki}\hat{n}_i, \qquad (168)$$

$$\hat{n}_i = \sum_k \hat{e}_{ki}\hat{m}_k. \qquad (169)$$

The stochastic collision operator can then be implemented independently for each mode m_k,

$$\hat{m}_k^{\star\text{neq}} = \gamma_k \hat{m}_k^{\text{neq}} + \varphi_k r_k, \qquad (170)$$

where r_k are independent Gaussian random variables with zero mean and unit variance. The dimensionless constants φ_k are determined by expressing the random stresses $\sigma_{\alpha\beta}^r$ in terms of the r_k and φ_k and then calculating the covariance matrix [102]. For example, $\sigma_{xy}^r = \sqrt{\mu\rho}c_s^2\varphi_7 r_7$, while $\sigma_{\alpha\alpha}^r = \sqrt{6\mu\rho}c_s^2\varphi_4 r_4$. Comparison with (163)–(165) shows that the correct stress correlations are obtained for

$$\varphi_k = \left(1 - \gamma_k^2\right)^{1/2}. \qquad (171)$$

At the hydrodynamic scale, only fluctuations in stress contribute to the time evolution of the momentum density (44) so in principle it is sufficient to add random fluctuations to the modes $m_4,\dots,m_9$ only: In the original derivation of the fluctuating LB equation [98, 100], the kinetic modes were projected out entirely, i.e., $\gamma_k = \varphi_k = 0$ for $k = 10, 11,\dots, 18$. More recently, Adhikari et al. [101] have argued that the kinetic modes should be thermalized as well. They extended (170) to the kinetic modes ($k = 10,\dots, 18$), with $\gamma_k = 0$ (as in [98, 100]) but with $\varphi_k = 1$, which then satisfies (171). It was demonstrated numerically that this leads to more accurate fluctuations at short length scales, but the theoretical justification remained somewhat obscure. From the discussion so far, we can see that both procedures give the same random stresses $\sigma_{\alpha\beta}^r$, and hence are not different from the point of view of fluctuating hydrodynamics. This has been clarified recently [102], by analyzing the LB model in terms of statistical mechanics. A purely microscopic approach was taken, in which the stochastic collisions were viewed as a Monte Carlo [116] process. Knowledge of the probability distribution of the LB variables **n** then makes it possible to check whether or not a given collision rule satisfies the condition of detailed balance. It can be shown [102] that the kinetic modes must be thermalized in order to satisfy detailed balance, in agreement with the procedure proposed in [101]. The theory will be outlined in Sect. 3.6.

3.6 Statistical Mechanics of Lattice-Boltzmann Models

The starting point of the statistical mechanical development in [102] is the notion
of a GLG. We define $v_i(\mathbf{r},t)$ in (30) as the *number* of particles with velocity $\mathbf{c}_i$ at
site $\mathbf{r}$ at time t. In contrast with the standard LB model, v_i is a (positive) integer;
in contrast with lattice-gas models, $v_i \gg 1$. The state at a particular lattice site,
$v(\mathbf{r},t)$, is modified by the collision process, subject to the constraints of mass and
momentum conservation; the post-collision state, $v^\star(\mathbf{r},t)$, is then propagated to the
neighboring sites (30).

Although a deterministic GLG collision operator would be difficult to construct,
we can nevertheless determine the distribution in a homogeneous equilibrium state
from the conservation laws alone. First we note that there is an entropy associated
with each v_i,

$$S_i = -(v_i \ln v_i - v_i - v_i \ln \bar{v}_i + \bar{v}_i), \tag{172}$$

where $\bar{v}_i$ is the mean value of v_i in the homogeneous state. Each velocity direction i
at each lattice point has a degeneracy $\exp(S_i)$, which can be derived from a Bernoulli
process. Particles are selected for the velocity direction i with probability p_0, with
p_0 chosen so that on average a total of $\bar{v} = N_r p_0$ particles will be selected from a
reservoir of N_r particles. Then the probability to select exactly v particles is given
by the binomial distribution,

$$p(v) = \frac{N_r!}{v!\,(N_r - v)!} \left(\frac{\bar{v}}{N_r}\right)^v \left(1 - \frac{\bar{v}}{N_r}\right)^{N_r - v}. \tag{173}$$

Equation (172) results from calculating $\ln p(v)$ in the limit of $N_r \to \infty$, at fixed $\bar{v}$.
Under the usual assumption that in the equilibrium state the populations corresponding to different lattice sites and different directions are uncorrelated, the entropy per
lattice site is $S(v) = \sum_i S_i$.

The populations at a given lattice site are sampled from a probability distribution
proportional to $\exp[S(v)]$, but subject to the constraints of fixed mass and momentum density, which characterize the homogeneous state:

$$P(v) \propto \exp[S(v)]\,\delta\left(\mu \sum_i v_i - \rho\right) \delta\left(\mu \sum_i v_i \mathbf{c}_i - \mathbf{j}\right). \tag{174}$$

Consistency with the formalism developed in the previous sections requires

$$\mu \bar{v}_i = \rho a^{c_i}. \tag{175}$$

The equilibrium or mean populations for a given ρ and $\mathbf{j}$ are found by maximizing
P or, more conveniently, by maximizing S and taking into account the conservation
laws by Lagrange multipliers:

$$\frac{\partial S}{\partial v_i} + \lambda_\rho + \lambda_{\mathbf{j}} \cdot \mathbf{c}_i = 0, \tag{176}$$

$$\mu \sum_i v_i - \rho = 0, \tag{177}$$

$$\mu \sum_i v_i \mathbf{c}_i - \mathbf{j} = 0. \tag{178}$$

The exact solution is

$$v_i^{\text{eq}} = \bar{v}_i \exp\left(\lambda_\rho + \lambda_{\mathbf{j}} \cdot \mathbf{c}_i\right), \tag{179}$$

where the Lagrange multipliers, λ_ρ and $\lambda_{\mathbf{j}}$, are found from the constraint equations (177) and (178). Solving these equations in terms of a power series in u, and disregarding terms of order $O(u^3)$, one finds the standard equilibrium distribution given in (97). This approach has been previously proposed within the framework of the "entropic lattice-Boltzmann" method [108, 109], which however, focuses exclusively on the deterministic LB model.

Within the statistical–mechanical framework we have developed for the LB model, the population densities n_i fluctuate around mean values determined by the hydrodynamic flow fields. Thus, the non-equilibrium distribution is sampled from $P(\mathbf{n}^{\text{neq}})$ which is Gaussian distributed about the equilibrium [102],

$$P(\mathbf{n}^{\text{neq}}) \propto \exp\left(-\sum_i \frac{(n_i^{\text{neq}})^2}{2\mu n_i^{\text{eq}}}\right) \delta\left(\sum_i n_i^{\text{neq}}\right) \delta\left(\sum_i \mathbf{c}_i n_i^{\text{neq}}\right). \tag{180}$$

The variance of the fluctuations is controlled by the mass density μ, associated with an LB particle. A small number of particles gives rise to large fluctuations and vice versa. For simplicity we will ignore the effects of flow on the variance of the distribution, replacing n_i^{eq} by its $u = 0$ value. This can be justified at the macroscopic level by the Chapman–Enskog expansion [102]. Rewriting (180) in terms of normalized variables $\hat{n}_i$ [see (166)], and transforming to the normalized modes [see (167) and (169)] eliminates the explicit constraints,

$$P(\hat{\mathbf{m}}^{\text{neq}}) \propto \exp\left(-\frac{1}{2} \sum_{k>3} \hat{m}_k^{\text{neq}\,2}\right). \tag{181}$$

Fluctuations only arise in the non-conserved modes, while the conserved modes have no non-equilibrium contribution, i.e., $m_k^{\text{neq}} = 0$ for $k \leq 3$.

We now reinterpret the update rule (170),

$$\hat{m}_k^{\star\text{neq}} = \gamma_k \hat{m}_k^{\text{neq}} + \varphi_k r_k, \tag{182}$$

as a Monte Carlo move. The transition probability is then identical to the probability of generating the random variable r_k,

$$\omega\left(\hat{m}_k^{\mathrm{neq}} \to \hat{m}_k^{\star\mathrm{neq}}\right) = \left(2\pi\varphi_k^2\right)^{-1/2} \exp\left[-\frac{\left(\hat{m}_k^{\star\mathrm{neq}} - \gamma_k\hat{m}_k^{\mathrm{neq}}\right)^2}{2\varphi_k^2}\right], \tag{183}$$

and for the reverse transition the same formula holds, with the pre- and post-collisional populations exchanged. The condition of detailed balance [116],

$$\frac{\omega\left(\hat{m}_k^{\mathrm{neq}} \to \hat{m}_k^{\star\mathrm{neq}}\right)}{\omega\left(\hat{m}_k^{\star\mathrm{neq}} \to \hat{m}_k^{\mathrm{neq}}\right)} = \frac{\exp\left[-(\hat{m}_k^{\star\mathrm{neq}})^2/2\right]}{\exp\left[-(\hat{m}^{\mathrm{neq}})^2/2\right]}, \tag{184}$$

then holds if and only if

$$\varphi_k = \left(1 - \gamma_k^2\right)^{1/2}, \tag{185}$$

as before (171). The important point is that this relation, which in the previous sub-section was only proved for the stress modes, can now be shown to hold for *all* non-conserved modes. It is a necessary condition for consistent sampling of the thermal fluctuations, not just on the macroscopic hydrodynamic level (for which the stress modes alone are sufficient), but also on the microscopic LB level itself. Although assigning $\gamma_k = 0$ (and $\varphi_k = 1$) to all kinetic modes is obvious and straight-forward [101], the present analysis shows that this is not necessary. Other values of γ_k and φ_k are possible as well, so long as they satisfy (185), and specific values may be desirable for a more accurate treatment of boundary conditions [113, 114].

3.7 External Forces

An external force density $\mathbf{f}(\mathbf{r},t)$ can be introduced into the LB algorithm by an additional collision operator Δ_i'',

$$\Delta_i = \sum_j \mathcal{L}_{ij}\left(n_j - n_j^{\mathrm{eq}}\right) + \Delta_i''. \tag{186}$$

For simplicity, we only consider the deterministic case; the analysis of the fluctu-ating part (Δ_i') remains the same. Application of the new collision operator should leave the mass density unchanged, but increase the momentum density by $h\mathbf{f}$. This implies the following conditions on the moments of Δ_i'':

$$\sum_i \Delta_i'' = 0, \tag{187}$$

$$\sum_i \Delta_i'' \mathbf{c}_i = h\mathbf{f}. \tag{188}$$

Consequently the definition of the fluid velocity is no longer unique: one can le-gitimately choose any value for $\mathbf{u}$ between $\rho^{-1}\left(\sum_i n_i\mathbf{c}_i\right)$ and $\rho^{-1}\left(\sum_i n_i\mathbf{c}_i + h\mathbf{f}\right)$ (i.e., between the pre-collisional and post-collisional states). However, numerical [98] and theoretical [100, 117, 118] analysis shows that the optimum value is just the arithmetic mean of the pre- and post-collisional velocities. We *define* the momentum density as

$$\mathbf{j} = \sum_i n_i \mathbf{c}_i + \frac{h}{2} \mathbf{f}, \tag{189}$$

and the corresponding flow velocity as $\mathbf{u} = \mathbf{j}/\rho$. Consistency with (66) requires that we use this value for $\mathbf{u}$ to calculate n_i^{eq} (97):

$$\sum_i n_i^{\mathrm{eq}} \mathbf{c}_i = \mathbf{j}, \tag{190}$$

$$\sum_i n_i^{\mathrm{neq}} \mathbf{c}_i = -\frac{h}{2} \mathbf{f}. \tag{191}$$

In [100] the usual moment condition $\sum_i n_i^{\mathrm{neq}} \mathbf{c}_i = 0$ was maintained. In comparison with the present approach this makes a small error of order f^2 to the distribution, which leads to spurious terms in the Chapman–Enskog analysis. In contrast the present approach leads to a clean result, entirely equivalent to the force-free case. This may be of consequence when there are strongly inhomogeneous forces, such as are considered in Sect. 4. However, it should also be noted that these differences vanish in the low Reynolds number limit.

Since $\Delta_i''(\mathbf{n}^{\mathrm{eq}}) \neq 0$, the Chapman–Enskog expansion of Δ_i'' starts at order ε^1 [cf. (56)]:

$$\Delta_i'' = \varepsilon \Delta_i''^{(1)} + \varepsilon^2 \Delta_i''^{(2)} + \cdots. \tag{192}$$

Following the procedure of Sect. 3.2, we take moments of (57) and (59) and obtain similar equations for the mass and momentum density [cf. (74) and (75)]:

$$\partial_t \rho + \partial_\alpha j_\alpha = 0, \tag{193}$$

$$\partial_t j_\alpha + \partial_\beta \pi_{\alpha\beta}^{\mathrm{eq}} + \frac{1}{2} \partial_\beta \left(\pi_{\alpha\beta}^{\star\mathrm{neq}} + \pi_{\alpha\beta}^{\mathrm{neq}} \right) = f_\alpha. \tag{194}$$

However, the second moment leads to a force-dependent contribution to the non-equilibrium momentum flux, which can be derived as before, beginning with (76) and substituting the equilibrium expressions for $\pi_{\alpha\beta}^{\mathrm{eq}}$ (79) and $\Phi_{\alpha\beta\gamma}^{\mathrm{eq}}$ (98). The time derivative of the momentum flux now generates terms involving $\mathbf{uf} + \mathbf{fu}$ from the momentum conservation equation on the t_1 time scale [cf. (99)]:

$$\pi_{\alpha\beta}^{\star\mathrm{neq}} - \pi_{\alpha\beta}^{\mathrm{neq}} = h\rho c_s^2 \left(\partial_\alpha u_\beta + \partial_\beta u_\alpha \right) + h \left(u_\alpha f_\beta + u_\beta f_\alpha \right). \tag{195}$$

Spurious terms proportional to $\mathbf{uf}$ can be eliminated from (194), by including the second moment of Δ_i'',

$$\Sigma_{\alpha\beta} = \frac{1}{h} \sum_i \Delta_i'' c_{i\alpha} c_{i\beta}, \tag{196}$$

so that the stress update is now [cf. (135) and (136)]

$$\pi_{\alpha\beta}^{\star neq} = \gamma_s \bar{\pi}_{\alpha\beta}^{neq} + \frac{1}{3} \gamma_v \pi_{\gamma\gamma}^{neq} \delta_{\alpha\beta} + h\Sigma_{\alpha\beta}. \tag{197}$$

Equations (195) and (197) form a linear system for $\pi_{\alpha\beta}^{neq}$ and $\pi_{\alpha\beta}^{\star neq}$. Solving these equations as before (139)–(142), and inserting the result into (194), we obtain a Newtonian stress with unchanged values for the viscosities by choosing $\Sigma_{\alpha\beta}$ such that

$$\Sigma_{\alpha\beta} = \frac{1}{2}(1+\gamma_s)\left[u_\alpha f_\beta + u_\beta f_\alpha - \frac{2}{3}u_\gamma f_\gamma \delta_{\alpha\beta}\right] + \frac{1}{3}(1+\gamma_v)u_\gamma f_\gamma \delta_{\alpha\beta}. \tag{198}$$

The moment conditions expressed by (187), (188) and (198) are uniquely satisfied by the choice

$$\Delta_i'' = a^{c_i}\left[\frac{h}{c_s^2}f_\alpha c_{i\alpha} + \frac{h}{2c_s^4}\Sigma_{\alpha\beta}\left(c_{i\alpha}c_{i\beta} - c_s^2\delta_{\alpha\beta}\right)\right], \tag{199}$$

where Δ_i'' only affects the modes $m_1, \ldots, m_9$. This result has been derived previously [118] within the context of the LBGK model; here we have presented the derivation in the more general MRT framework.

4 Coupling the LB Fluid to Soft Matter

The fundamental algorithmic problem in soft matter simulations is the coupling between the solid and fluid phases. A key attraction of LB methods is the simplicity with which geometrically complex boundaries can be incorporated. The first correct implementation of a moving boundary condition was described in the proceedings of a workshop on LGCA [119]; a more accessible source is [120]. The idea was to modify the bounce-back rule for stationary surfaces such that the steady-state distribution was consistent with the local surface velocity. By constructing the boundary-node interactions along the individual links, the viscous stress remains unchanged. Subsequently, we showed numerically that this algorithm gives accurate hydrodynamic interactions between spherical particles suspended in a lattice-gas fluid [120]. Nevertheless, it quickly became clear that the fluctuating LB model was a more useful computational tool, for the reasons outlined in Sect. 3. The LG algorithm for the moving boundary condition carries over in a simple and direct way to the LB method [98]. A number of improvements to the bounce-back boundary condition have been proposed over the years, and we will summarize some of the more practical approaches in Sect. 4.4.

More recently an entirely different approach has been proposed [121, 122] in which particles couple to the fluid through a frictional drag. This method has the advantage of greatly reducing the number of LB grid points in the simulation, at the cost of a representation that is only correct in the far field. The method has been applied to polymers [43, 122, 123] and to suspended solid particles [124–128]. In the latter case the surface is described by a number of sources distributed over the surface of the particle. The distributed forces resemble the Immersed Boundary (IB) methods [129], which are common in finite-difference and finite-element simulations; this connection has only been recognized recently [130]. We will summarize these developments and add some new ideas and interpretation of the force coupling methods. In a related development [131, 132], conventional immersed boundary methods are being used in conjunction with an LB fluid. However, the coupling in this case is implicit, solving for the velocity of the interface through a force balance, which corresponds to the high friction limit of [124–128]. Here, we will only consider inertial coupling, since the theory for thermal fluctuations has not been worked out for the implicit schemes.

4.1 Boundary Conditions

To simulate the hydrodynamic interactions between solid particles in suspension, the LB model must be modified to incorporate the boundary conditions imposed on the fluid by the solid particles. The basic methodology is illustrated in Fig. 1. The solid particles are defined by a boundary surface, which can be of any size or shape; in Fig. 1 it is a circle. When placed on the lattice, the boundary surface cuts some of the links between lattice nodes. The fluid particles moving along these links interact

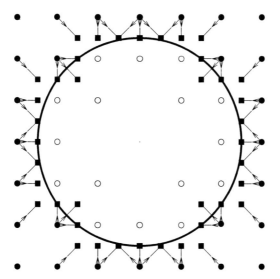

Fig. 1 Location of boundary nodes for a curved surface. The velocities along links cutting the boundary surface are indicated by *arrows*. The locations of the boundary nodes are shown by *solid squares*, and the fluid nodes by *solid circles*. The *open circles* indicate nodes in the solid adjacent to fluid nodes

with the solid surface at boundary nodes placed halfway along the links. Thus, a discrete representation of the particle surface is obtained, which becomes more and more precise as the particle gets larger.

In early work, the lattice nodes on either side of the boundary surface were treated in an identical fashion [98, 120], so that fluid filled the whole volume of space, both inside and outside the solid particles. Although the fluid motion inside the particle closely follows that of a rigid solid body [99], at short times the inertial lag of the fluid is noticeable, and the contribution of the interior fluid to the particle force and torque reduces the stability of the particle velocity update. Today, most simulations exclude interior fluid, although the implementation is more difficult when the particles move. The moving boundary condition [98] without interior fluid [133] is then implemented as follows. We take the set of fluid nodes $\mathbf{r}$ just outside the particle surface, and for each node all the velocities $\mathbf{c}_b$ such that $\mathbf{r} + \mathbf{c}_b h$ lies inside the particle surface. An example of a set of boundary node velocities is shown by the arrows in Fig. 1. Each of the corresponding population densities is then updated according to a simple rule which takes into account the motion of the particle surface [98];

$$n_{b'}(\mathbf{r}, t + h) = n_b^*(\mathbf{r}, t) - \frac{2a^{c_b} \rho \mathbf{u}_b \cdot \mathbf{c}_b}{c_s^2}, \tag{200}$$

where $n_b^*(\mathbf{r}, t)$ is the post-collision distribution at $(\mathbf{r}, t)$ in the direction $\mathbf{c}_b$, and $\mathbf{c}_{b'} = -\mathbf{c}_b$. The local velocity of the particle surface,

$$\mathbf{u}_b = \mathbf{U} + \mathbf{\Omega} \times (\mathbf{r}_b - \mathbf{R}), \tag{201}$$

is determined by the particle velocity $\mathbf{U}$, angular velocity $\mathbf{\Omega}$, and center of mass $\mathbf{R}$; $\mathbf{r}_b = \mathbf{r} + \frac{1}{2} h \mathbf{c}_b$ is the location of the boundary node.

As a result of the boundary node updates, momentum is exchanged locally between the fluid and the solid particle, but the combined momentum of solid and fluid is conserved. The forces exerted at the boundary nodes can be calculated from the momentum transferred in (200), and the particle forces and torques are then obtained by summing over all the boundary nodes associated with a particular particle. It can be shown analytically that the force on a planar wall in a linear shear flow is exact [98], and several numerical examples of LB simulations of hydrodynamic interactions are given in [99]. Figure 2 illustrates the accuracy that can be achieved with the MRT collision operator described in Sect. 3.4. Even with small particles, only $5b$ in diameter, the hydrodynamic interactions are within 1% of a precise numerical solution [21], down to separations between the particle surfaces $s = r - 2a \sim b$, corresponding to $s \sim 0.4a$, where a is the sphere radius. Periodic boundaries with a unit cell size $L = 12a$ were used, with the pair inclined at 30° to a symmetry axis; other geometries give a very similar level of agreement. We emphasize that there are no adjustable parameters in these comparisons. In particular, in contrast to previous work [99, 104], there is no need to calibrate the particle radius; the correct particle size arises automatically when the eigenvalues of the

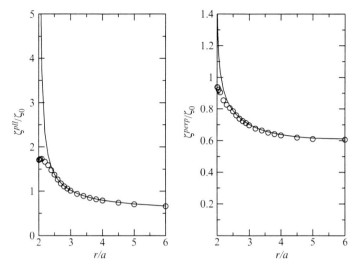

Fig. 2 Hydrodynamic interactions from LB simulations with particles of radius $a = 2.5b$. The *solid symbols* are the LB friction coefficients, ζ^{pll} and ζ^{perp}, for the relative motion of two spheres along the line of centers (*left*) and perpendicular to the line of centers (*right*). Results are compared with essentially exact results from a multipole code [21] in the same geometry (*solid lines*)

kinetic modes of the MRT model have the appropriate dependence on the shear viscosity [114].

To understand the physics of the moving boundary condition, one can imagine an ensemble of particles, moving at constant speed $\mathbf{c}_b$, impinging on a massive wall oriented perpendicular to the particle motion. The wall itself is moving with velocity $\mathbf{u}_b \ll \mathbf{c}_b$. The velocity of the particles after collision with the wall is $-\mathbf{c}_b + 2\mathbf{u}_b$ and the force exerted on the wall is proportional to $\mathbf{c}_b - \mathbf{u}_b$. Since the velocities in the LB model are discrete, the desired boundary condition cannot be implemented directly, but we can instead modify the density of returning particles so that the momentum transferred to the wall is the same as in the continuous velocity case. It can be seen that this implementation of the no-slip boundary condition leads to a small mass transfer across a moving solid–fluid interface. This is physically correct and arises from the discrete motion of the solid surface. Thus, during a time step h the fluid is flowing continuously, while the solid particle is fixed in space. If the fluid cannot flow across the surface there will be large artificial pressure gradients, arising from the compression and expansion of fluid near the surface. For a uniformly moving particle, it is straightforward to show that the mass transfer across the surface in a time step h (200) is exactly recovered when the particle moves to its new position. For example, each fluid node adjacent to a planar wall has five links intersecting the wall. If the wall is advancing into the fluid with a velocity $\mathbf{U}$, then the mass flux across the interface (from 200) is $\rho \mathbf{U}$. Apart from small compressibility effects, this is exactly the rate at which fluid mass is absorbed by the moving wall. For sliding motion, (200) correctly predicts no net mass transfer across the interface.

4.2 Particle Motion

An explicit update of the particle velocity

$$\mathbf{U}(t+h) = \mathbf{U}(t) + \frac{h}{m}\mathbf{F}(t) \tag{202}$$

has been found to be unstable [99] unless the particle radius is large or the particle mass density is much higher than the surrounding fluid. In previous work [99] the instability was reduced, but not eliminated, by averaging the forces and torques over two successive time steps. Subsequently, an implicit update of the particle velocity was proposed [134] as a means of ensuring stability. A generalized version of that idea, which can be adapted to situations where two particles are in near contact, was developed in [104]. Here we sketch an elaboration of this idea, which is consistent with a Trotter decomposition of the Liouville operator [135–139]. We will only consider the update of the position and linear velocity explicitly; the extension to rotational motion is straightforward [104].

The equations of motion for the suspended particles are written as

$$\dot{\mathbf{R}}_i = \mathbf{U}_i, \tag{203}$$
$$m\dot{\mathbf{U}}_i = \mathbf{F}_i^h(\mathbf{R}_i, \mathbf{U}_i) + \mathbf{F}_i^c(\mathbf{R}^N), \tag{204}$$

where we have separated the forces into a hydrodynamic component $\mathbf{F}_i^h$, which depends on the particle position and velocity, and a conservative force $\mathbf{F}_i^c$, which depends on the positions of all particles. The hydrodynamic force depends on the fluid degrees of freedom as well, but these remain unchanged during the particle update and need not be considered as dynamical variables here.

A second-order Trotter decomposition [135–139] breaks the update of a single time step into three independent components: a half-time step update of the positions at constant velocity, a full-time step update of the velocities with fixed positions, and a further half time step update of the positions using the new velocities:

$$\mathbf{R}_i(t+\tfrac{1}{2}h) = \mathbf{R}_i(t) + \frac{h}{2}\mathbf{U}_i(t), \tag{205}$$

$$\dot{\mathbf{U}}_i = \frac{1}{m}\left[\mathbf{F}_i^h(\mathbf{R}_i(t+\tfrac{1}{2}h), \mathbf{U}_i) + \mathbf{F}_i^c(\mathbf{R}^N(t+\tfrac{1}{2}h))\right], \tag{206}$$

$$\mathbf{R}_i(t+h) = \mathbf{R}_i(t+\tfrac{1}{2}h) + \frac{h}{2}\mathbf{U}_i(t+h). \tag{207}$$

In the absence of velocity-dependent forces this is just the Verlet scheme, but the solid–fluid boundary conditions (200) introduce a hydrodynamic force that depends linearly on the particle velocity [104, 134],

$$\mathbf{F}_i^h(\mathbf{R}_i, \mathbf{U}_i) = \mathbf{F}_0^h(\mathbf{R}_i) - \zeta(\mathbf{R}_i) \cdot \mathbf{U}_i. \tag{208}$$

The velocity independent force is calculated at the half-time step

$$\mathbf{F}_0^h(\mathbf{R}_i(t + \tfrac{1}{2}h)) = \frac{b^3}{h} \sum_b 2n_b^*(\mathbf{r}, t)\mathbf{c}_b, \tag{209}$$

where the sum is over all the boundary nodes, b, describing the particle surface and $\mathbf{c}_b$ points towards the particle center. The location of the boundary nodes is determined by the particle coordinates $\mathbf{R}_i(t + \tfrac{1}{2}h)$, which should be evaluated at the half-time step as indicated. The post-collision populations, n_b^*, are calculated at time t but arrive at the boundary nodes at the half-time step also. The components of the matrix

$$\zeta(\mathbf{R}_i(t + \tfrac{1}{2}h)) = \frac{2\rho b^3}{c_s^2 h} \sum_b a^{cb} \mathbf{c}_b \mathbf{c}_b \tag{210}$$

are high-frequency friction coefficients, which describe the instantaneous force on a particle in response to a sudden change in velocity. Complete expressions, including rotation, are given in [104].

The LB fluid and the solid particles are coupled by an instantaneous momentum transfer at the half-time step, which is therefore presumed to be conservative:

$$\mathbf{U}_i(t + \tfrac{1}{2}h) = \mathbf{U}_i(t) + \frac{h}{2m} \mathbf{F}_i^c(\mathbf{R}^N(t + \tfrac{1}{2}h)), \tag{211}$$

$$\mathbf{U}_i^*(t + \tfrac{1}{2}h) = \mathbf{U}_i(t + \tfrac{1}{2}h) + \frac{h}{m} \mathbf{F}_i^h(\mathbf{R}_i(t + \tfrac{1}{2}h), \tilde{\mathbf{U}}_i(t + \tfrac{1}{2}h)), \tag{212}$$

$$\mathbf{U}_i(t + h) = \mathbf{U}_i^*(t + \tfrac{1}{2}h) + \frac{h}{2m} \mathbf{F}_i^c(\mathbf{R}^N(t + \tfrac{1}{2}h)). \tag{213}$$

However, it is not entirely clear what velocity should be used in (212): among the possibilities discussed in [104] are an explicit update $\tilde{\mathbf{U}}_i(t + \tfrac{1}{2}h) = \mathbf{U}_i(t + \tfrac{1}{2}h)$, an implicit update $\tilde{\mathbf{U}}_i(t + \tfrac{1}{2}h) = \mathbf{U}_i^*(t + \tfrac{1}{2}h)$, and a semi-implicit update $\tilde{\mathbf{U}}_i(t + \tfrac{1}{2}h) = [\mathbf{U}_i(t + \tfrac{1}{2}h) + \mathbf{U}_i^*(t + \tfrac{1}{2}h)]/2$. It has been pointed out [140] that, even for a Langevin equation with constant friction, there are deviations in the temperature for finite values of h. However, the semi-implicit scheme satisfies the FDT exactly for constant friction. Here we will consider a different model for the velocity, assuming that the hydrodynamic force is distributed over the time step. For simplicity we consider a single component of the velocity,

$$m\dot{U} = -\zeta U + F_0^h + F^c, \tag{214}$$

where ζ, F_0^h and F^c are constant in this context. The solution of (214) over a time interval h is

$$U(t + h) = U(t)\exp(-\alpha) + \frac{F_0^h + F^c}{\zeta}[1 - \exp(-\alpha)], \tag{215}$$

where $\alpha = \zeta h/m$ is the dimensionless time step. Equation (215) is stable for all values of α, satisfies the FDT exactly, and, when there is no conservative force, encompasses previous algorithms as limiting cases. Both explicit [98] and implicit [104, 134] schemes are consistent with an expansion of (215) to linear order in α, while the semi-implicit method [104] can be derived from a second-order expansion in α. The steady-state velocity $U(t+h) = U(t)$ satisfies the force balance $F^h + F^c = 0$ exactly. This new result may lead to more accurate integration of the particle positions and velocities in the large α limit.

To complete the update, the velocity $\tilde{U}_i(t + \frac{1}{2}h)$ is needed to calculate the momentum transfer to the fluid (200). An explicit update [98] can be done in a single pass since $\tilde{U}_i(t + \frac{1}{2}h) = U_i(t)$ is already known, but an implicit or semi-implicit update requires two passes through the boundary nodes. The first pass is used to calculate F_0^h so that (212) can be solved for $\tilde{U}_i(t + \frac{1}{2}h)$ [104]. This velocity is used to update the population densities in a second sweep through the boundary nodes. In the present case we calculate $\tilde{U}_i(t + \frac{1}{2}h)$ by enforcing consistency between the sequential update (211)–(213) and (215):

$$\alpha \tilde{U}_i(t + \tfrac{1}{2}h) = U_i(t)\left[1 - \exp(-\alpha)\right] + \left(F_0^h + F^c\right)\left[\frac{\alpha}{\zeta} - \frac{1 - \exp(-\alpha)}{\zeta}\right]. \quad (216)$$

This ensures overall momentum conservation as before.

When there are short-range conservative forces between the particles, the LB time step is frequently too large for accurate integration of the interparticle forces. The LB time step can be divided into an integer number of substeps, but the question then arises as to how to best incorporate the hydrodynamic forces, since F^h should, in principle, be calculated at $t + \frac{1}{2}h$. One possibility is to use the fact that ζ varies slowly with particle position and accept the small error associated with using $R(t)$ rather than $R(t + \frac{1}{2}h)$. Or this solution could be used as a predictor step for calculating $R(t + \frac{1}{2}h)$, which could then be followed by one or more corrector cycles with increasingly more accurate calculations of $\zeta(t + \frac{1}{2}h)$. The corrector cycles should not involve a significant overhead since the boundary nodes would be largely the same from one cycle to the next, and the time-consuming lookup of LB population densities could be avoided.

Although the momentum exchange between fluid and solid occurs instantaneously at the half time step, in calculating $\tilde{U}_i(t + \frac{1}{2}h)$ we have made the assumption that the hydrodynamic force is distributed over the time step. We actually attempted to derive an update for the velocity assuming that the hydrodynamic force acts over a very small fraction of the time step, but this has not led to a sensible result as yet. It is not entirely clear if the assumption that the hydrodynamic force acts over the whole time step is valid, and does not, for example, produce an artificial dissipation. To resolve this question will require a detailed analysis of the fully coupled system, along the lines given in Sect. 4.5 for the simpler case of frictional coupling. A similar analysis for solid–fluid boundary conditions is an open area for further research.

4.3 Surfaces Near Contact

When two particle surfaces come within one grid spacing, fluid nodes are excluded from regions between the solid surfaces, leading to a loss of mass conservation. This happens because boundary updates at each link cause mass transfer across the solid–fluid interface, which is necessary to accommodate the discrete motion of the particle surface (see Sect. 4.1). The total mass transfer in or out of an isolated particle is

$$\Delta M = -\frac{2h^3\rho}{c_s^2}\left[\mathbf{U}\cdot\sum_b a^{C_b}\mathbf{c}_b\right] = 0, \tag{217}$$

regardless of the particle's size or shape.

Although the sum $\sum_b a^{C_b}\mathbf{c}_b$ is zero for any closed surface [104], when two particles are close to contact some of the boundary nodes are missing and the surfaces are no longer closed. In this case $\Delta M \neq 0$ and mass conservation is no longer ensured. Two particles that remain in close proximity never reach a steady state, no matter how slowly they move, since fluid is constantly being added or removed, depending on the particle positions and velocities. If the two particles move as a rigid body mass conservation is restored, but in general this is not the case. The accumulation or loss of mass occurs slowly, and in many dynamical simulations it fluctuates with changing particle configuration but shows no long-term drift. However, we typically enforce mass conservation, particle-by-particle, by redistributing the excess mass among the boundary nodes [104]. An alternative idea is to ensure that there is always at least one fluid node in the gap between the particle surfaces. In dense suspensions it would be quite inaccurate to insert an artificial excluded volume around the particles, but a more promising idea is to cut back the particle surfaces along planes perpendicular to the line of centers [141, 142]. It remains to be seen if the hydrodynamic interactions retain the level of accuracy shown in Fig. 2.

When two particles are in near contact, the fluid flow in the gap cannot be resolved. For particle sizes that are typically used in multiparticle simulations ($a < 5b$), the lubrication breakdown in the calculation of the hydrodynamic interaction occurs at gaps of the order of $0.1a$. However, in some flows, notably the shearing of a dense suspension, qualitatively important physics occurs at smaller separations, typically down to $0.01a$. Here we outline a method to implement lubrication corrections into a LB simulation.

For particles close to contact, the lubrication force, torque, and stresslet can be calculated from a sum of pairwise-additive contributions [29], and if we consider only singular terms, they can be calculated from the particle velocities alone [143]. In LB simulations [100, 144] the calculated forces follow the Stokes flow results down to a fixed separation, approximately equal to the grid spacing b, and remain roughly constant thereafter (see Fig. 2). The simplest lubrication correction is to take the difference between the lubrication force at a gap s and the force at some cut off distance s_c; i.e.,

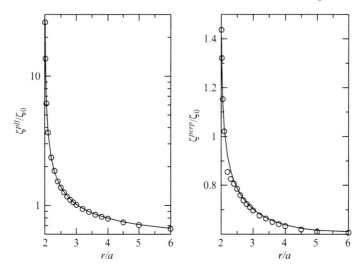

Fig. 3 Hydrodynamic interactions including lubrication, with particles of radius $a = 2.5b$. The *solid symbols* are the LB friction coefficients, ζ^{pll} and ζ^{perp}, for the relative motion of two spheres along the line of centers (*left*) and perpendicular to the line of centers (*right*). Results are compared with essentially exact results from a multipole code [21] in the same geometry (*solid lines*)

$$\mathbf{F}^l = -6\pi\eta \frac{a_1^2 a_2^2}{(a_1 + a_2)^2} \left(\frac{1}{s} - \frac{1}{s_c} \right) \mathbf{U}_{12} \cdot \hat{\mathbf{R}}_{12} \hat{\mathbf{R}}_{12}, \quad s < s_c \qquad (218)$$

$$\mathbf{F}^l = 0, \quad s > s_c, \qquad (219)$$

where $\mathbf{U}_{12} = \mathbf{U}_1 - \mathbf{U}_2$, $s = |\mathbf{R}_{12}| - a_1 - a_2$ is the gap between the two surfaces, and the unit vector $\hat{\mathbf{R}}_{12} = \mathbf{R}_{12}/|\mathbf{R}_{12}|$. Numerical tests of this procedure for the older 10-moment LB model are reported in [104]. Results for the MRT model are shown in Fig. 3, using a cutoff distance $s_c = 1.1b$ for the parallel component and $s_c = 0.7b$ for the perpendicular component. Even this simple form for the correction gives an accurate description of the lubrication regime, with the largest deviations occurring near the patch points. A more accurate correction can be obtained by calibrating each distance separately as in Stokesian dynamics and related methods [21, 145].

4.4 Improvements to the Bounce-Back Boundary Condition

The bounce-back boundary condition remains the most popular choice for simulations of suspensions, because of its robustness and simplicity. The results in Figs. 2 and 3 show that accurate hydrodynamic interactions, within 1–2%, can be achieved with quite small particles, particularly when combined with the MRT model. The reason that bounce-back works so well, despite being only first-order accurate, is that the errors in the momentum transfer tend to cancel when averaged over a

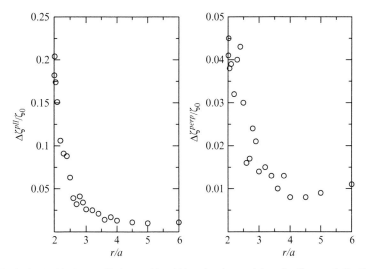

Fig. 4 Variation in friction coefficients with grid location for particles of radius $a = 2.5b$. The *solid symbols* are the variance in the LB friction coefficients, $\Delta\zeta^{\text{pll}}$ and $\Delta\zeta^{\text{perp}}$ for the relative motion of two spheres along the line of centers (*left*) and perpendicular to the line of centers (*right*). Results are a single standard deviation in the friction coefficients, calculated from 100 independent positions with respect to the grid

random sampling of boundary node positions [114]. In fact bounce-back can sometimes be more accurate than interpolation, where the errors, though locally smaller, do not cancel.

The most important deficiency of the bounce-back algorithm is the dependence of the force on the position of the nodes with respect to the grid. The results in Figs. 2 and 3 are averages over 100 independent configurations, in which the relative positions of the particles are the same but the pair is displaced randomly with respect to the underlying lattice. However, the variance in the friction for randomly sampled grid locations is small, typically of the order of 1%, as can be seen in Fig. 4. Nevertheless there is a much larger fluctuation in the force around the particle surface, which is particularly problematic if the particles are deformable [146, 147]. Thus, while the bounce-back method is quite accurate on average, locally the errors can be large. A detailed analytical and numerical critique of the bounce-back algorithm can be found in [148], together with an analysis of several of the modifications mentioned below. The most practical higher-order boundary conditions are adapted from the link bounce-back algorithm outlined in Sect. 4.1.

More sophisticated boundary conditions have been developed using finite-volume methods [149, 150] and interpolation [151–153]. A simple, physically motivated interpolation scheme has been proposed [151, 154], which both improves the accuracy of the bounce-back rule and is unconditionally stable for all boundary positions; the scheme has both linear and quadratic versions. A more general framework for this class of interpolation schemes has been extensively analyzed in a comprehensive and seminal paper [113]; the Multi-Reflection Rule proposed in [113] is the most accurate boundary condition yet discovered for LB methods. However, interpolation requires additional fluid nodes in the gap between adjacent particle surfaces. The bounce-back rule requires only one grid point between the surfaces but linear interpolation requires at least two grid points, while quadratic interpolation and multi-reflection require three. Recently, it was proposed that only the equilibrium distribution needs to be interpolated [114]. Although this is more complex to implement than linear interpolation, it has the advantage that the velocity distribution at the boundary surface may be used to provide an additional interpolation point. In this way the span of fluid nodes can be reduced to that of the bounce-back rule, while obtaining second-order accuracy in the flow field. In conjunction with an appropriate choice of collision operator [113], the location of the hydrodynamic boundary remains independent of fluid viscosity, unlike the linear and quadratic interpolations [151]. For viscous fluids, where $\gamma_s > 0$, the equilibrium interpolation rule is more accurate than either linear or quadratic interpolation [114].

4.5 Force Coupling

The force-coupling algorithm [121, 122] starts from a system of mass points which are coupled *dissipatively* to the hydrodynamic continuum. The particles are specified by positions $\mathbf{r}_i$, momenta $\mathbf{p}_i$, masses m_i, and phenomenological friction coefficients Γ_i. They interact via a potential $V\left(\mathbf{r}^N\right)$, giving rise to conservative forces $\mathbf{F}_i^c = -\partial V/\partial \mathbf{r}_i$. The fluid exerts a drag force on each particle based on the difference between the particle velocity and the fluid velocity $\mathbf{u}_i = \mathbf{u}(\mathbf{r}_i)$,

$$\mathbf{F}_i^d = -\Gamma_i \left(\frac{\mathbf{p}_i}{m_i} - \mathbf{u}_i \right). \tag{220}$$

Momentum conservation requires that an equal and opposite force be applied to the fluid. Both discrete and continuous degrees of freedom are subject to Langevin noise in order to balance the frictional and viscous losses, and thereby keep the temperature constant. The algorithm can be applied to any Navier–Stokes solver, not just to LB models. For this reason, we will discuss the coupling within a (continuum) Navier–Stokes framework, with a general equation of state $p(\rho)$. We use the abbreviations $\eta_{\alpha\beta\gamma\delta}$ for the viscosity tensor (46), and

$$\pi_{\alpha\beta}^E = p\delta_{\alpha\beta} + \rho u_\alpha u_\beta \tag{221}$$

for the inviscid momentum flux or Euler stress (79). Since the fluid equations are solved on a grid, whereas the particles move continuously, it will be necessary to *interpolate* the flow field from nearby lattice sites to the particle positions [122].

The addition of a point force into the continuum fluid equations introduces a singularity into the flow field, which causes both mathematical and numerical difficulties. On the other hand, the flow field around a finite-sized particle can be generated by a distributed force located entirely inside the particle [155, 156]. This flow field is everywhere finite, and the force density appearing in (17) can be written as

$$\mathbf{f}(\mathbf{r}) = -\sum_i \mathbf{F}_i^d \Delta(\mathbf{r}, \mathbf{r}_i), \tag{222}$$

where $\Delta(\mathbf{r})$ is a weight function with compact support and normalization

$$\int d^3\mathbf{r} \Delta(\mathbf{r}, \mathbf{r}_i) = 1. \tag{223}$$

Compact support limits the set of nodes $\mathbf{r}$ to those in the vicinity of $\mathbf{r}_i$ and ensures that the interactions remain local. Away from solid boundaries, translational invariance requires that

$$\Delta(\mathbf{r}, \mathbf{r}_i) = \Delta(\mathbf{r} - \mathbf{r}_i). \tag{224}$$

The function $\Delta(\mathbf{r}, \mathbf{r}_i)$ plays a dual role, both interpolating the fluid velocity field to the particle position,

$$\mathbf{u}(\mathbf{r}_i) = \int d^3\mathbf{r} \Delta(\mathbf{r}, \mathbf{r}_i) \mathbf{u}(\mathbf{r}), \tag{225}$$

and then redistributing the reactive force to the fluid, according to (222). Within the context of polymer simulations, Δ has been regarded as an interpolating function for point forces, but it can equally well be regarded as a model for a specific distributed force, contained within an envelope described by $\Delta(\mathbf{r} - \mathbf{r}_i)$. The flow fields from a point force and a distributed force are similar at large distances from the source, but the distributed source has the advantage that the near field also corresponds to a physical system, namely finite-size particles. We will adopt the distributed source interpretation both here and in Sect. 4.6.

The Langevin equations of motion for the coupled fluid–particle system are

$$\frac{d}{dt}\mathbf{r}_i = \frac{1}{m_i}\mathbf{p}_i, \tag{226}$$

$$\frac{d}{dt}\mathbf{p}_i = \mathbf{F}_i^c + \mathbf{F}_i^d + \mathbf{F}_i^f, \tag{227}$$

$$\partial_t \rho + \partial_\alpha j_\alpha = 0, \tag{228}$$

$$\partial_t j_\alpha + \partial_\beta \pi_{\alpha\beta}^E = \partial_\beta \eta_{\alpha\beta\gamma\delta} \partial_\gamma u_\delta + f_\alpha^h + \partial_\beta \sigma_{\alpha\beta}^f, \tag{229}$$

where the force density applied to the fluid includes both dissipative and random forces,

$$\mathbf{f}^h(\mathbf{r}) = -\sum_i \left(\mathbf{F}_i^d + \mathbf{F}_i^f \right) \Delta(\mathbf{r}, \mathbf{r}_i). \tag{230}$$

The Langevin noises for the particles and fluid, $\mathbf{F}_i^f$ and $\sigma_{\alpha\beta}^f$, satisfy the usual moment conditions:

$$\left\langle F_{i\alpha}^f \right\rangle = 0, \tag{231}$$

$$\left\langle \sigma_{\alpha\beta}^{f} \right\rangle = 0, \tag{232}$$

$$\left\langle F_{i\alpha}^{f}(t) F_{j\beta}^{f}(t') \right\rangle = 2k_{B} T \Gamma_{i} \delta_{ij} \delta_{\alpha\beta} \delta\left(t - t'\right), \tag{233}$$

$$\left\langle \sigma_{\alpha\beta}^{f}(\mathbf{r},t) \sigma_{\gamma\delta}^{f}(\mathbf{r}',t') \right\rangle = 2k_{B} T \eta_{\alpha\beta\gamma\delta} \delta\left(\mathbf{r} - \mathbf{r}'\right) \delta\left(t - t'\right). \tag{234}$$

By construction, this coupling is local, and conserves both the total mass

$$M = \sum_{i} m_{i} + \int d^{3}r \rho \tag{235}$$

and the total momentum

$$\mathbf{P} = \sum_{i} \mathbf{p}_{i} + \int d^{3}r \rho \mathbf{u}. \tag{236}$$

Galilean invariance is ensured by using velocity differences in the coupling between particles and fluid (220). A finer point is that the interpolation uses $\mathbf{u}$ (and not $\mathbf{j}$), so that the velocity field enters strictly linearly. We will prove that the fluctuation–dissipation theorem (FDT) holds for this coupled system, proceeding in three steps that successively take more terms into account.

Let us look first at the conservative system where the particles and fluid are completely decoupled

$$\frac{d}{dt}\mathbf{r}_{i} = \frac{1}{m_{i}}\mathbf{p}_{i}, \tag{237}$$

$$\frac{d}{dt}\mathbf{p}_{i} = \mathbf{F}_{i}^{c}, \tag{238}$$

$$\partial_{t}\rho + \partial_{\alpha} j_{\alpha} = 0, \tag{239}$$

$$\partial_{t} j_{\alpha} + \partial_{\beta} \pi_{\alpha\beta}^{E} = 0. \tag{240}$$

The dynamics of the particles *and* the Euler fluid can be described within the framework of Hamiltonian mechanics [157]. The Hamiltonians for the particles

$$\mathcal{H}_{p} = \sum_{i} \frac{\mathbf{p}_{i}^{2}}{2m_{i}} + V, \tag{241}$$

and fluid,

$$\mathcal{H}_{f} = \int d^{3}\mathbf{r} \left(\frac{1}{2}\rho \mathbf{u}^{2} + \epsilon\left(\rho\right) \right), \tag{242}$$

are conserved quantities, with $\epsilon(\rho)$ the internal energy density of the fluid.

As a second step, we consider a system where particles and fluid are still decoupled, but are subject to dissipation and noise:

$$\frac{d}{dt}\mathbf{r}_{i} = \frac{1}{m_{i}}\mathbf{p}_{i}, \tag{243}$$

$$\frac{d}{dt}\mathbf{p}_i = \mathbf{F}_i^c - \frac{\Gamma_i}{m_i}\mathbf{p}_i + \mathbf{F}_i^f, \tag{244}$$

$$\partial_t \rho + \partial_\alpha j_\alpha = 0, \tag{245}$$

$$\partial_t j_\alpha + \partial_\beta \pi_{\alpha\beta}^E = \partial_\beta \eta_{\alpha\beta\gamma\delta}\partial_\gamma u_\delta + \partial_\beta \sigma_{\alpha\beta}^f. \tag{246}$$

These Langevin equations are known to satisfy the FDT [15, 42, 158–160]. We briefly sketch the formalism used for the proof, since this will be needed for the final step in which we consider the fully coupled system.

Instead of describing the stochastic dynamics via a Langevin equation, we use the Fokker–Planck equation, which is the evolution equation for the probability density in phase space. For an N-particle system,

$$\partial_t P\left(\mathbf{r}^N, \mathbf{p}^N\right) = (\mathcal{L}_1 + \mathcal{L}_2 + \mathcal{L}_3) P\left(\mathbf{r}^N, \mathbf{p}^N\right), \tag{247}$$

where $\mathbf{r}^N, \mathbf{p}^N$ denote the positions and momenta of all N particles. The three operators $\mathcal{L}_1$, $\mathcal{L}_2$, and $\mathcal{L}_3$ describe the Hamiltonian, frictional, and stochastic part of the dynamics; they can be found via the Kramers–Moyal expansion [15, 159]:

$$\mathcal{L}_1 = -\sum_i \left(\frac{\partial}{\partial \mathbf{r}_i} \cdot \frac{\mathbf{p}_i}{m_i} + \frac{\partial}{\partial \mathbf{p}_i} \cdot \mathbf{F}_i^c\right), \tag{248}$$

$$\mathcal{L}_2 = \sum_i \frac{\Gamma_i}{m_i}\frac{\partial}{\partial \mathbf{p}_i} \cdot \mathbf{p}_i, \tag{249}$$

$$\mathcal{L}_3 = k_B T \sum_i \Gamma_i \frac{\partial^2}{\partial \mathbf{p}_i^2}. \tag{250}$$

The FDT holds if the Boltzmann factor, $\exp(-\mathcal{H}_p/k_B T)$, is a stationary solution of the Fokker–Planck equation. Using $\beta = (k_B T)^{-1}$ to define the inverse temperature, we have

$$\mathcal{L}_1 \exp(-\beta\mathcal{H}_p) = 0 \tag{251}$$

as a direct consequence of energy conservation in Hamiltonian systems. Furthermore, the relation

$$(\mathcal{L}_2 + \mathcal{L}_3)\exp(-\beta\mathcal{H}_p) = 0 \tag{252}$$

can be shown by direct differentiation.

For the fluid system, the phase space comprises all possible configurations of the fields $\rho(\mathbf{r})$, $\mathbf{j}(\mathbf{r})$, which we denote as $[\rho]$, $[\mathbf{j}]$. The Fokker–Planck equation for the fluid degrees of freedom can be written as

$$\partial_t P([\rho], [\mathbf{j}]) = (\mathcal{L}_4 + \mathcal{L}_5 + \mathcal{L}_6) P([\rho], [\mathbf{j}]), \tag{253}$$

where $\mathcal{L}_4$, $\mathcal{L}_5$, and $\mathcal{L}_6$ describe the Hamiltonian, viscous, and stochastic components,

$$\mathcal{L}_4 = \int d^3\mathbf{r} \left(\frac{\delta}{\delta\rho} \partial_\alpha j_\alpha + \frac{\delta}{\delta j_\alpha} \partial_\beta \pi^E_{\alpha\beta} \right), \tag{254}$$

$$\mathcal{L}_5 = -\eta_{\alpha\beta\gamma\delta} \int d^3\mathbf{r} \frac{\delta}{\delta j_\alpha} \partial_\beta \partial_\gamma u_\delta, \tag{255}$$

$$\mathcal{L}_6 = k_B T \eta_{\alpha\beta\gamma\delta} \int d^3\mathbf{r} \int d^3\mathbf{r}' \frac{\delta}{\delta j_\alpha(\mathbf{r})} \frac{\delta}{\delta j_\gamma(\mathbf{r}')} \left[\frac{\partial}{\partial r_\beta} \frac{\partial}{\partial r'_\delta} \delta(\mathbf{r} - \mathbf{r}') \right], \tag{256}$$

and $\delta \ldots / \delta \ldots$ represents a functional derivative [161, see, e.g.,]. Replacing $\partial/\partial r'_\delta$ in the last equation with $-\partial/\partial r_\delta$ enables integration over $\mathbf{r}'$:

$$\mathcal{L}_6 = -k_B T \eta_{\alpha\beta\gamma\delta} \int d^3\mathbf{r} \frac{\delta}{\delta j_\alpha} \partial_\beta \partial_\gamma \frac{\delta}{\delta j_\delta}, \tag{257}$$

where we have exploited the symmetry of the viscosity tensor with respect to the indexes γ and δ. Functional differentiation of the Boltzmann factor with respect to $\mathbf{j}$,

$$\frac{\delta}{\delta j_\delta} \exp\left(-\beta\mathcal{H}_f\right) = -\beta u_\delta \exp\left(-\beta\mathcal{H}_f\right), \tag{258}$$

then shows that

$$(\mathcal{L}_5 + \mathcal{L}_6) \exp\left(-\beta\mathcal{H}_f\right) = 0. \tag{259}$$

Finally, the relation

$$\mathcal{L}_4 \exp\left(-\beta\mathcal{H}_f\right) = 0 \tag{260}$$

follows from energy conservation in Hamiltonian dynamics.

We now turn to the coupled system, with Hamiltonian $\mathcal{H} = \mathcal{H}_p + \mathcal{H}_f$. The Fokker–Planck equation in the full phase space reads

$$\partial_t P\left(\mathbf{r}^N, \mathbf{p}^N, [\rho], [\mathbf{j}]\right) = \left(\sum_{i=1}^{10} \mathcal{L}_i\right) P\left(\mathbf{r}^N, \mathbf{p}^N, [\rho], [\mathbf{j}]\right), \tag{261}$$

with the operators $\mathcal{L}_7$–$\mathcal{L}_{10}$ to describe the coupling in the equations of motion:

$$\mathcal{L}_7 = -\sum_i \Gamma_i \frac{\partial}{\partial p_{i\alpha}} u_{i\alpha}, \tag{262}$$

$$\mathcal{L}_8 = -\sum_i \Gamma_i \int d^3\mathbf{r} \Delta(\mathbf{r}, \mathbf{r}_i) \frac{\delta}{\delta j_\alpha(\mathbf{r})} \left(\frac{1}{m_i} p_{i\alpha} - u_{i\alpha} \right), \tag{263}$$

$$\mathcal{L}_9 = k_B T \sum_i \Gamma_i \int d^3\mathbf{r} \Delta(\mathbf{r}, \mathbf{r}_i) \frac{\delta}{\delta j_\alpha(\mathbf{r})} \int d^3\mathbf{r}' \Delta(\mathbf{r}', \mathbf{r}_i) \frac{\delta}{\delta j_\alpha(\mathbf{r}')}, \tag{264}$$

$$\mathcal{L}_{10} = -2k_B T \sum_i \Gamma_i \frac{\partial}{\partial p_{i\alpha}} \int d^3\mathbf{r} \Delta(\mathbf{r}, \mathbf{r}_i) \frac{\delta}{\delta j_\alpha(\mathbf{r})}. \tag{265}$$

The coupling of the fluid velocity to the particles is described by $\mathcal{L}_7$, while $\mathcal{L}_8$ describes the drag on the fluid. The stochastic contributions include fluid–fluid correlations via $\mathcal{L}_9$, and fluid–particle cross correlations via $\mathcal{L}_{10}$.

It should be noted that $\mathbf{u}_i$, being the result of the interpolation, depends on the fields $[\rho]$ and $[\mathbf{j}]$, so that $\delta u_i / \delta j_\alpha(\mathbf{r})$ is nonzero. Hence, the corresponding operators in $\mathcal{L}_8$ do not commute. Explicit functional differentiation shows that (258) holds in an analogous way for the interpolated velocity,

$$\int d^3 \mathbf{r} \Delta(\mathbf{r}, \mathbf{r}_i) \frac{\delta}{\delta j_\alpha(\mathbf{r})} \exp(-\beta \mathcal{H}) = -\beta u_{i\alpha} \exp(-\beta \mathcal{H}). \tag{266}$$

Explicit calculations, as outlined for the uncoupled system, show that

$$(\mathcal{L}_7 + \mathcal{L}_8 + \mathcal{L}_9 + \mathcal{L}_{10}) \exp(-\beta \mathcal{H}) = 0, \tag{267}$$

which implies the exact FDT for the fully coupled system,

$$\left(\sum_{i=1}^{10} \mathcal{L}_i \right) \exp(-\beta \mathcal{H}) = 0. \tag{268}$$

An important consequence of this result is that a consistent simulation needs to thermalize *both* fluid and particle degrees of freedom; any other choice will violate the FDT.

The Langevin integrator for the particles is constructed in much the same way as the velocity Verlet algorithm for MD. Although a Langevin analog to the Verlet algorithm has been known for some time [162], straightforward derivations have become available only recently, by applying operator-splitting techniques that were previously limited to Hamiltonian systems [163]. We employ a second-order integrator [135–137], which reduces to the velocity Verlet scheme in the limit of vanishing friction. Higher-order schemes are known [164], but they are considerably more complicated. Specifically, we approximately integrate the equations

$$\frac{d}{dt} \mathbf{r}_i = \frac{1}{m_i} \mathbf{p}_i, \tag{269}$$

$$\frac{d}{dt} \mathbf{p}_i = \mathbf{F}_i^c + \mathbf{F}_i^d + \mathbf{F}_i^f, \tag{270}$$

assuming that the fluid velocity $\mathbf{u}_i$ is constant over a time step h. This corresponds to the Fokker–Planck equation for the particles

$$\partial_t P(\mathbf{r}^N, \mathbf{p}^N, t) = (\mathcal{L}_r + \mathcal{L}_p) P(\mathbf{r}^N, \mathbf{p}^N, t) \tag{271}$$

with

$$\mathcal{L}_r = -\sum_i \frac{\partial}{\partial \mathbf{r}_i} \cdot \frac{\mathbf{p}_i}{m_i}, \tag{272}$$

$$\mathcal{L}_p = -\sum_i \frac{\partial}{\partial \mathbf{p}_i} \cdot \mathbf{F}_i^c + \sum_i \Gamma_i \frac{\partial}{\partial \mathbf{p}_i} \cdot \left(\frac{\mathbf{p}_i}{m_i} - \mathbf{u}_i \right) + k_B T \sum_i \Gamma_i \frac{\partial^2}{\partial \mathbf{p}_i^2}. \tag{273}$$

The formal solution

$$P\left(\mathbf{r}^N, \mathbf{p}^N, h\right) = \exp\left[\left(\mathcal{L}_r + \mathcal{L}_p\right) h\right] P\left(\mathbf{r}^N, \mathbf{p}^N, 0\right) \tag{274}$$

is approximated by a second-order Trotter decomposition,

$$\exp\left[\left(\mathcal{L}_r + \mathcal{L}_p\right) h\right] = \exp\left(\mathcal{L}_r h/2\right) \exp\left(\mathcal{L}_p h\right) \exp\left(\mathcal{L}_r h/2\right) + O(h^3). \tag{275}$$

The operator splitting implies the following algorithm: a half-time step update of the coordinates with *constant momenta*,

$$\mathbf{r}_i(t + h/2) = \mathbf{r}_i(t) + \frac{h}{2} \frac{\mathbf{p}_i(t)}{m_i}, \tag{276}$$

followed by a full-time step momentum update, with *constant coordinates*,

$$\mathbf{p}_i(t + h) = C_i^{(1)}(h) \mathbf{p}_i(t) + C_i^{(2)}(h) \left[\mathbf{F}_i^c + \Gamma_i \mathbf{u}_i\right] + C_i^{(3)}(h) \theta_i, \tag{277}$$

and finally another half-time step coordinate update, with *constant momenta*,

$$\mathbf{r}_i(t + h) = \mathbf{r}_i(t + h/2) + \frac{h}{2} \frac{\mathbf{p}_i(t + h)}{m_i}. \tag{278}$$

The coefficients in (277) are

$$C_i^{(1)}(h) = \exp\left(-\frac{\Gamma_i}{m_i} h\right), \tag{279}$$

$$C_i^{(2)}(h) = \frac{m_i}{\Gamma_i} \left[1 - \exp\left(-\frac{\Gamma_i}{m_i} h\right)\right], \tag{280}$$

$$C_i^{(3)}(h) = \sqrt{m_i k_B T \left[1 - \exp\left(-2\frac{\Gamma_i}{m_i} h\right)\right]}, \tag{281}$$

where $\theta_{i\alpha}$ are Gaussian random variables with zero mean and unit variance. Equation (277) is the *exact* solution of the momentum update, since (270) is a linear Langevin equation describing Brownian motion in a harmonic potential [158]. Thus, the only source of error in integrating the particle motion is derived from the Trotter decomposition itself, (275). Nevertheless, it is important to limit the range of random numbers, to ensure that very large steps do not occasionally occur. It is therefore both desirable and more efficient to use distributions of random variates with finite range, which reproduce the Gaussian moments up to a certain order; in the present case, $O(h^2)$ accuracy requires that moments up to the fourth cumulant are correct. One possible choice is

$$P(\theta) = \frac{2}{3}\delta(\theta) + \frac{1}{6}\delta\left(\theta - \sqrt{3}\right) + \frac{1}{6}\delta\left(\theta + \sqrt{3}\right). \tag{282}$$

After the update of the particle momenta, the fluid force density, $\mathbf{f}^h(\mathbf{r})$ (230), is distributed to the surrounding lattice sites. After one or more particle updates, the fluid variables are updated for a single LB time step, with the external forces being taken into account in the collision operator. This scheme is probably only first-order accurate overall. It remains a challenge for the future to develop a unified framework to describe the fully coupled system and analyze its convergence properties; the algorithm could then perhaps be improved in a systematic fashion.

The input friction coefficient is *not* the same as the long-time friction coefficient, which is measured by the ratio of the particle velocity to the applied force [122]. Consider an isolated particle with "bare" (or input) friction coefficient Γ dragged through the fluid by a constant force $\mathbf{F}$, resulting in a steady particle velocity $\mathbf{U}$. The force balance requires that

$$\mathbf{F} = -\mathbf{F}^d = \Gamma(\mathbf{U} - \mathbf{u}_0), \tag{283}$$

where $\mathbf{u}_0$ is the fluid velocity at the particle center, $\mathbf{r}_0$,

$$\mathbf{u}_0 = \int d^3\mathbf{r}\Delta(\mathbf{r}, \mathbf{r}_0)\mathbf{u}(\mathbf{r}). \tag{284}$$

In the absence of thermal fluctuations, the deterministic fluid velocity field can be calculated in the Stokes flow approximation using the Green's function appropriate to the boundary conditions [22],

$$\mathbf{u}(\mathbf{r}) = \int d^3\mathbf{r}' \mathsf{T}(\mathbf{r}, \mathbf{r}') \cdot \mathbf{F}\Delta(\mathbf{r}', \mathbf{r}_0). \tag{285}$$

In an unbounded fluid, the Green's function reduces to the Oseen tensor [22], $\mathsf{T}(\mathbf{r}, \mathbf{r}') = \mathsf{O}(\mathbf{r} - \mathbf{r}')$, with

$$\mathsf{O}(\mathbf{r}) = \frac{1}{8\pi\eta r}\left(1 + \frac{\mathbf{r}\mathbf{r}}{r^2}\right), \tag{286}$$

but Green's functions are also known for periodic boundary conditions [165] and planar boundaries [38, 39] as well. Combining (283), (284) and (285),

$$\mathbf{u}_0 = \mathsf{T}^{av} \cdot \mathbf{F} = \Gamma\mathsf{T}^{av} \cdot (\mathbf{U} - \mathbf{u}_0), \tag{287}$$

where T^{av} is the Green's function averaged over the particle envelope,

$$T^{av}(\mathbf{r}_0) = \int d^3\mathbf{r} \int d^3\mathbf{r}' \Delta(\mathbf{r},\mathbf{r}_0)\Delta(\mathbf{r}',\mathbf{r}_0)T(\mathbf{r},\mathbf{r}'). \tag{288}$$

In a system with translational invariance, T^{av} is independent of $\mathbf{r}_0$ and proportional to the unit tensor by symmetry. Then from (287)

$$\mathbf{F} = \frac{\Gamma}{1 + \mu_\infty \Gamma}\mathbf{U}, \tag{289}$$

where $\mu_\infty = T^{av}_{\alpha\alpha}/3$ accounts for the renormalization of Γ. If the size of the particle a is associated with the range of interaction of Δ, then dimensional analysis of (288) suggests that $\mu_\infty \sim (\eta a)^{-1}$. The effective friction coefficient, defined via $\mathbf{F} = \Gamma_{eff}\mathbf{U}$, is therefore diminished by the flow field induced by the applied force,

$$\frac{1}{\Gamma_{eff}} = \frac{1}{\Gamma} + \mu_\infty. \tag{290}$$

This relation has been verified numerically by extensive computer experiments [122, 140]. In the strong coupling limit, $\Gamma \to \infty$, the effective friction saturates to the limiting value $\Gamma_{eff} = \mu_\infty^{-1}$. This suggests assigning an effective radius to the interpolating function,

$$\frac{1}{a} = 6\pi\eta\mu_\infty. \tag{291}$$

However, for smaller values of the input friction, the effective particle size is given by

$$\frac{1}{a} = 6\pi\eta\left(\frac{1}{\Gamma} + \mu_\infty\right) = \frac{1}{a_0} + \frac{1}{gb}, \tag{292}$$

where $a_0 = \Gamma/6\pi\eta$ is the input particle radius and $gb = (6\pi\eta\mu_\infty)^{-1}$ depends on the interpolating function. The interesting physical parameter is Γ_{eff}, which describes the long-time behavior of the coupled system. Thus, to approach the continuum limit, one should keep Γ_{eff} constant as the lattice spacing is decreased, and change the bare coupling Γ as necessary.

It is not yet known how $T^{av}(\mathbf{r}_0)$ behaves in the vicinity of a solid boundary, when translational invariance is broken. The compact support of the weighting function Δ suggests that T^{av} is a local correction and therefore largely independent of macroscopic boundary conditions. Numerical simulations with periodic boundary conditions (Sect. 4.6) show that g is independent of system size and fluid viscosity. The weak system-size dependence reported in [140] is entirely accounted for by the difference between the periodic Green's function [165] and the Oseen tensor. Thus, in a periodic unit cell of length L, (292) requires a correction of order $1/L$ [19],

$$\frac{1}{gb} = \frac{1}{a} + \frac{2.84}{L} - \frac{1}{a_0}. \tag{293}$$

4.6 Interpolating Functions

In this section we consider translationally invariant interpolating functions, $\Delta(\mathbf{r}, \mathbf{r}_0)$ $= \Delta(\mathbf{r} - \mathbf{r}_0)$ in more detail. For a discrete lattice, the interpolation procedure [cf. (225)] reads

$$\mathbf{u}(\mathbf{r}_0) = \sum_{\mathbf{r}} \Delta(\mathbf{r} - \mathbf{r}_0)\mathbf{u}(\mathbf{r}), \tag{294}$$

where $\mathbf{r}_0$ is the position of the particle, and $\mathbf{r}$ denotes the lattice sites. The normalization condition

$$\sum_{\mathbf{r}} \Delta(\mathbf{r} - \mathbf{r}_0) = 1 \tag{295}$$

must hold for all particle positions, $\mathbf{r}_0$, in order for the conservation laws to be satisfied exactly. We will assume the analysis of Sect. 4.5 carries over to the discrete system with no more than second-order discretization errors. Proof of this assumption remains for future work; here we numerically compare various choices of Δ.

Previous work, incorporating force coupling into spectral codes, used an isotropic Gaussian distribution for $\Delta(r)$ [155], but this is not commensurate with cubic lattice symmetry. Thus, we take Δ as a product of one-dimensional functions [129]

$$\Delta(x, y, z) = \phi\left(\frac{x}{b}\right) \phi\left(\frac{y}{b}\right) \phi\left(\frac{z}{b}\right). \tag{296}$$

We first consider the two-point linear interpolating polynomial

$$\phi_2(u) = \begin{cases} 1 - |u| & |u| \leq 1, \\ 0 & |u| \geq 1, \end{cases} \tag{297}$$

which satisfies the following moment conditions for all real-valued u and integer j:

$$\sum_{j} \phi(u - j) = 1, \tag{298}$$

$$\sum_{j} j\phi(u - j) = u. \tag{299}$$

It exactly conserves momentum and angular momentum of the particle and fluid. However, ϕ_2 violates the condition

$$\sum_{j} \phi^2(u - j) = C, \tag{300}$$

where C is a constant, independent of u. The importance of this condition is explained in [129].

The conditions in (298)–(300) can be satisfied by a three-point interpolation function,

$$\phi_3(u) = \begin{cases} \frac{1}{3}\left(1+\sqrt{1-3u^2}\right) & 0 \le |u| \le \frac{1}{2}, \\ \frac{1}{6}\left(5-3|u|-\sqrt{-2+6|u|-3u^2}\right) & \frac{1}{2} \le |u| \le \frac{3}{2}, \\ 0 & \frac{3}{2} \le |u|. \end{cases} \tag{301}$$

An important property that emerges from ϕ_3 is that the first derivative, $\phi_3'(u)$, is continuous throughout the whole domain of u. This ensures that the velocity field varies smoothly across the grid, with a continuous spatial derivative $\nabla \mathbf{u}$. By contrast, linear interpolation leads to a continuous velocity but discontinuous derivatives.

In order to test the various interpolation schemes we have determined the settling velocity of a single particle in a periodic unit cell. A small force was applied to the particle and a compensating pressure gradient (or uniform force density) was added to the fluid, so that the net force on the system was zero. The steady-state particle velocity was determined, without allowing the particle to move on the grid [99]. This procedure is valid in Stokes flow, where an arbitrarily small velocity may be assumed, and gives a clean result for the variation in settling velocity with grid position. We used (293) to convert the measured mobility to a single parameter g, which does not depend on system size (L) or fluid viscosity (η). The mobility of the particle is reduced by the periodic images [165], but the correction in (293) accounts for the effects of the periodic boundaries quantitatively [99, 120, 166]. In simulations of polymer solutions an average of g over all grid positions is used.

The smoother velocity field derived from three-point interpolation means that the particle velocity is less dependent on the underlying grid than with linear interpolation, as can be seen in Fig. 5. Here we show the variation in effective

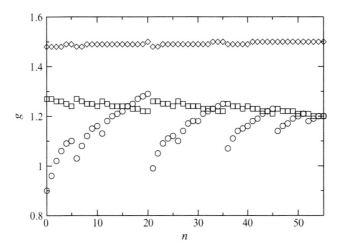

Fig. 5 Variation in settling velocity with grid location. The effective hydrodynamic radius was determined from $U = F/6\pi\eta a$ and converted to g using (293); g was found to be independent of η and L, as expected. Results are shown at 56 different grid positions (labeled by the index n), systematically varying the coordinates in steps of $0.1b$. Particles with input radius $a_0 = b$ were placed at coordinates $(ib/10, jb/10, kb/10)$, with $0 \le i \le j \le k \le 5$. Results are shown for the two-point (*circles*), three-point (*squares*) and four-point (*diamonds*) interpolation schemes

particle size, as determined by the parameter g (293), for linear (two point) interpolation (297), three-point interpolation (301), and the four-point interpolation,

$$\phi_4(u) = \begin{cases} \frac{1}{8}\left(3 - 2|u| + \sqrt{1 + 4|u| - 4u^2}\right) & 0 \le |u| \le 1, \\ \frac{1}{8}\left(5 - 2|u| - \sqrt{-7 + 12|u| - 4u^2}\right) & 1 \le |u| \le 2, \\ 0 & 2 \le |u|, \end{cases} \tag{302}$$

that is commonly used in immersed boundary methods [129]. As was recently noticed [130], the four-point interpolation leads to a much smaller variation in effective friction than linear interpolation. The parameter g varies with grid position by up to 20% in the case of linear interpolation, but by less than 1% with four-point interpolation. On the other hand, linear interpolation requires an envelope volume of eight grid points, while the four-point scheme requires 64 grid points. Away from the strong-coupling limit, the grid dependence of the settling velocity is reduced since there is a non-negligible input mobility apart from the lattice contribution.

Four-point interpolation is only necessary when using centered-difference approximations to the velocity and pressure fields [129], a situation that does not arise in LB simulations. We see that the three-point scheme is also much smoother than linear interpolation, with about a 3% variation in g. It is not as smooth as the four-point interpolation, but requires only 27 grid points. Furthermore, the smaller span of nodes means that the boundary surface is more tightly localized, and in fact the hydrodynamic interactions obtained with three-point interpolations are just as accurate as those obtained with four-point interpolation, as shown below.

An important test of the force coupling scheme is its ability to represent the hydrodynamic interactions between two spherical particles. As an example of the accuracy of the different interpolation schemes, in Fig. 6 we show the hydrodynamic interactions between two spheres moving along the line of centers. A small force is applied to sphere 1, in the direction of the vector between 1 and 2, and the velocity of sphere 2 is determined. From this we can calculate the hydrodynamic mobility μ_{12}^{pll}. The results are normalized by the mobility of the isolated sphere $\mu_0 = (6\pi\eta a)^{-1}$. Results were obtained for the two-point, three-point, and four-point interpolation schemes, using a source particle placed on a grid point $(0,0,0)$ in one instance and in the center of the voxel $(b/2, b/2, b/2)$ in the other. The simulations were carried out in a periodic unit cell, with 20 grid points in each direction. Results are compared with a spectral solution of the Stokes equations in a periodic geometry, assuming the force density on the particle surface is constant. For an isolated pair of particles this level of approximation includes the Oseen interaction and the Faxen correction; it corresponds to the Rotne–Prager (RP) interaction [23] used in most Brownian dynamics simulations of hydrodynamically interacting particles:

$$\mathbf{U}_2 = \left[\mathsf{T}(\mathbf{R}_{12}) + \frac{a^2}{12\pi\eta R_{12}^5}\left(R_{12}^2 \mathbf{1} - 3\mathbf{R}_{12}\mathbf{R}_{12}\right)\right] \cdot \mathbf{F}_1. \tag{303}$$

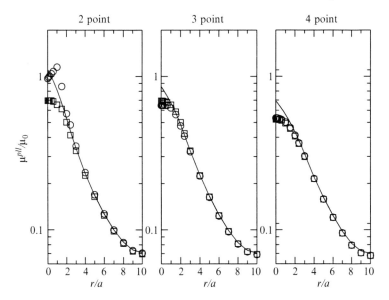

Fig. 6 Hydrodynamic interactions between a pair of spherical particles using the force-coupling method. The normalized mobility $\mu_{12}^{p\parallel}/\mu_0$ is plotted at various separations r. We fit the effective hydrodynamic radius of each interpolating function to numerical solutions of the Stokes equations with a *uniform* force density on the sphere. These results correspond to a Rotne–Prager description of the hydrodynamic mobility and do not include lubrication. We show results at two different grid locations, $(0,0,0)$ (*circles*) and $(b/2, b/2, b/2)$ (*squares*), for two-point (*left*), three-point (*center*) and four-point (*right*) interpolation schemes

The periodic RP tensor can be calculated by Ewald summation [167] or by direct summation of Fourier components, which converges if used in conjunction with finite-volume sources [17, 168].

The results for the two-point interpolation show a significant dependence on the exact grid position when the particles are close to each other, $r < 4a$, but the results for the higher-order schemes are essentially independent of the grid. The three-point integration scheme is of comparable accuracy to the four-point scheme but requires less than half the number of grid points. It would seem to be the best choice for applications even though the particle motion is not quite as smooth. When the particles are widely separated, the simulations match almost perfectly with both the Oseen and RP solutions for the same periodic geometry; the typical errors are of the order of 0.1% of the Stokes velocity.

When the spheres are closer together, $r < 3a$, then the simulated hydrodynamic interactions match the Rotne–Prager interaction rather better than the Oseen interaction. This confirms that the weight function does make the particles behave as volume sources, rather than points. The best fit between simulation results and Stokes flow is obtained for an effective particle radius that is roughly $0.33w$, where w is the range of the weight function. So for two-point interpolation ($w = 2b$) the effective size is about $0.7b$, for three-point interpolation ($w = 3b$) it is about $1.0b$, and for four-point interpolation ($w = 4b$) it is about $1.3b$. The actual values used in Fig. 6 are $0.8b$, $1.0b$, and $1.2b$, respectively. The optimal hydrodynamic radius is quite large, $a \sim b$, corresponding to strong coupling, $a_0 \sim 5b$. It remains an open question whether it is practical or desirable to run the fluctuating simulations with such large input friction.

5 Applications with Hydrodynamic Interactions

In this section we will discuss applications of the LB method to simulations of soft matter. We will briefly summarize some of our published work in this area, with the aim of indicating the breadth of possible applications of the method. Results will be summarized from simulations that cover a wide range of experimental length and time scales – from nanometers to millimeters and from nanoseconds to seconds.

5.1 Short-Time Diffusion of Colloids

The first application of the fluctuating LB model was to the short-time diffusion of hard-sphere colloids. At the time, a new experimental technique – Diffusing Wave Spectroscopy (DWS) [169] – enabled the study of the dynamics of colloids on time scales of a few nanoseconds. In general, the diffusion of a colloidal particle is a Markov process, but at such short times the developing hydrodynamic flow field gives rise to additional long-range correlations, analogous to the "long-time tails" in MD [9]. Although the existence of long-time tails had been established theoretically [170,171] and by MD simulations [9], these experiments marked the first direct observation of correlated hydrodynamic fluctuations. Brownian and Stokesian dynamics both neglect long-range dynamic correlations, using instead the Stokes-flow approximation, which is typically only valid on time scales longer then $1\,\mu s$.

Long-time tails occur naturally in the dynamics of lattice-gas models of colloidal suspensions [172] and even of the lattice gases themselves [173]. It might be supposed that such correlations would be absent in a Boltzmann-level model, due to the Stosszahlansatz closure assumption. The fluctuating LB model described in Sect. 3 does not have any long-time tails in the stress autocorrelation functions, but mode coupling between the diffusion of fluid momentum and the diffusion of the colloidal particle does lead to an algebraic decay of the velocity correlation function of a suspended sphere [174]. In these simulations the particle-fluid coupling was implemented via the link-bounce-back (BB) algorithm [98, 120] described in Sect. 4.1. Figure 7 shows the decay of translational and rotational velocity from two different types of computer experiment. In one case an initial velocity is imposed, which decays away due to viscous dissipation, and in the other the particle is set in motion by stress fluctuations in the fluid. Figure 7 shows that, within statistical errors, the normalized velocity correlation functions are identical to the steady

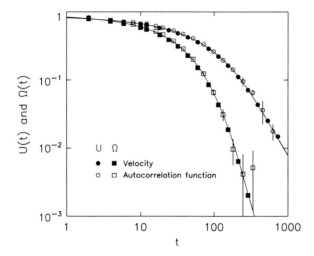

Fig. 7 Decay of translational (U) and rotational (Ω) velocity correlations of a suspended sphere. The time-dependent velocities of the sphere are shown as *solid symbols*; the relaxation of the corresponding velocity autocorrelation functions are shown as *open symbols* (with statistical error bars). A sufficiently large fluid volume was used so that the periodic boundary conditions had no effect on the numerical results for times up to $t = 1,000$ in lattice units ($h = b = 1$). The *solid lines* are theoretical results, obtained by an inverse Laplace transform of the frequency-dependent friction coefficients [175] of a sphere of appropriate size ($a = 2.6$) and mass ($\rho_s/\rho = 12$); the kinematic viscosity of the pure fluid $\eta_{\text{kin}} = 1/6$

decay of the translational and rotational velocities of the sphere; thus our simulations satisfy the fluctuation–dissipation theorem. Moreover, the simulations agree almost perfectly with theoretical results derived from the frequency-dependent friction coefficients [175], even though there are no adjustable parameters in these comparisons; thus we see that the fluctuating LB equation can account for the hydrodynamic memory effects that lead to long-time tails [9].

The simulations were also used to measure self diffusion in dense colloidal suspensions, up to a solids volume fraction of 45%. The simulation data, shown in Fig. 8, exhibits the same scaling with amplitude and time found in the DWS experiments [176]. In Fig. 8 the amplitude of the mean-square displacement has been normalized by its limiting value $6D_s t$, where D_s is the short-time self-diffusion coefficient. $D_s(\phi)$ is a monotonically decreasing function of concentration, because neighboring particles increasingly restrict the hydrodynamic flow field generated by the diffusing particle. Although the colloidal particles are freely moving in the fluid, the no-slip boundary condition induces stresslets and higher force multipoles on the particle surfaces; $D_s(\phi)$ is one average measure of these hydrodynamic interactions. The normalized mean-square displacement has a single relaxation time, so that when the time axis is scaled by τ all the data collapse onto a single curve, which is the same as for an isolated sphere. The relaxation time τ is the viscous diffusion time $\rho a^2/\eta$, of a single particle in a fluid of viscosity η, where $\eta(\phi)$ is numerically similar to the high-frequency viscosity of the suspension [19]. This

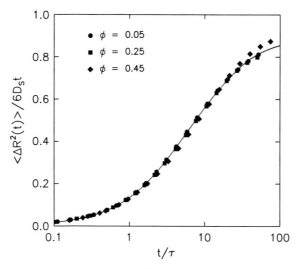

Fig. 8 Scaled mean-square displacement $\langle \Delta R^2(t) \rangle / 6D_s t$ at short times, vs. reduced time t/τ. Simulation results for 128 spheres (*solid symbols*) are shown at packing fractions ϕ of 5%, 25%, and 45%; the *solid line* is the isolated sphere result. The suspension viscosity at these packing fractions is $1.14\eta_0$, $2.17\eta_0$, and $5.6\eta_0$ respectively, where η_0 is the viscosity of the pure fluid

observation, which is in line with experimental measurements [176, 177], suggests that the short-time diffusion is essentially mean field like.

5.2 Dynamic Scaling in Polymer Solutions

The classical theory [6] of the equilibrium dynamics of polymer chains in solution is the Zimm model [178], which considers a single flexible chain in a good solvent, such that its conformations are given by a random coil with excluded volume segments:

$$R \sim bN^\nu. \tag{304}$$

Here, R is the size of the coil, measured in terms of the gyration radius or the end-to-end distance, while N denotes the number of monomers in the chain or the degree of polymerization, and b is the monomer size. Long-range interactions like electrostatics, or effects of poor solvent quality, are not considered. Furthermore, the solution is considered to be dilute, such that the chains do not overlap and a single-chain picture is sufficient. In other words, the standard Zimm model applies in the upper left corner of the generic phase diagram given in Fig. 9. Here the exponent ν takes the value $\nu \approx 0.59$ in three dimensions, because the excluded volume interaction leads to swelling of the chain when compared to an ideal random coil ($\nu = 1/2$). A polymer with excluded-volume is thus a self-similar random fractal with dimension $1/\nu$.

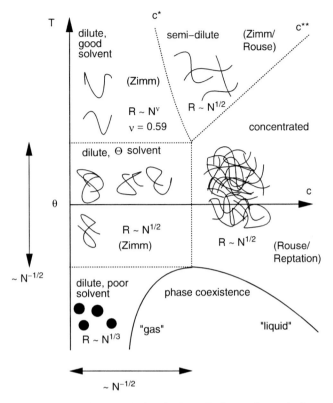

Fig. 9 Phase diagram of a polymer solution, in the c–T plane, where c is the monomer concentration and T is the temperature (parameterizing solvent quality). The static properties are characterized by scaling laws which describe the dependence of the chain size R (gyration radius or end-to-end-distance for example) on the degree of polymerization N. In the dilute limit ($c \rightarrow 0$) the so-called theta transition occurs, where at $T = \Theta$, single isolated chains collapse from a swollen random coil to a compact globule. For finite chain length N, this transition is "smeared out" over a temperature region $\Delta T \propto N^{-1/2}$, in which the chain conformations are Gaussian. Below Θ, there is phase coexistence between a "gas" of globules and a "liquid" of strongly interpenetrating Gaussian chains. The corresponding critical point occurs at a very low concentration, $c_c \propto N^{-1/2}$, and in the vicinity of Θ, $\Theta - T_c \propto N^{-1/2}$. The crossover region, which connects the regime of swollen isolated coils with that of the concentrated (Gaussian) solution at high temperatures, is called the semidilute regime. The dynamics is characterized by the Zimm model in the dilute limit where hydrodynamic interactions are important, and by the Rouse model for dense systems where they are screened. For very dense systems or sufficiently long chains, where curvilinear motion dominates, the Rouse model must be replaced by the reptation model (or the crossover behavior between these two cases). The Rouse and Zimm models are briefly described in the text

The Zimm model is based on the Rouse model [6, 179], but includes long-range hydrodynamic interactions between the segments. Both models predict self-similarity, not only with respect to space, but also with respect to time. Therefore, the dynamics is conveniently described in terms of an exponent z, connecting the chain relaxation time τ_R with the size of the coil R:

$$\tau_R \propto R^z. \tag{305}$$

The internal degrees of freedom completely re-organize on a time scale τ_R, leading to statistically independent conformations. This is also the time that the chain needs to diffuse through a distance equal to its own size:

$$D_{cm} \tau_R \sim R^2, \tag{306}$$

where D_{cm} is the translational diffusion coefficient, describing the center-of-mass motion. These two relations can be combined to determine the scaling of D_{cm} with chain size:

$$D_{cm} \propto R^{2-z}. \tag{307}$$

In the *Zimm* model, the hydrodynamic interactions result in strongly correlated motions, such that the coil as a whole behaves like a Stokes sphere, $D_{cm} \propto R^{-1}$, or

$$z = 3. \tag{308}$$

The *Rouse* model neglects hydrodynamic interactions and the monomer friction coefficients add up to give a total friction coefficient that is linearly proportional to N. Since $D_{cm} \propto N^{-1} \propto R^{-1/\nu}$,

$$z = 2 + \frac{1}{\nu}, \tag{309}$$

corresponding to slower dynamics.

Self-similarity implies that the relaxation of the internal degrees of freedom also scales with the exponent z on time scales $\tau_b \ll t \ll \tau_R$, where τ_b is the relaxation time on the monomer scale b. In the space–time window $b \ll l \ll R$, $\tau_b \ll t \ll \tau_R$, there is a scaling of the mean square displacement of a monomer,

$$\left\langle \Delta r^2 \right\rangle \propto t^{2/z}, \tag{310}$$

while the dynamic structure factor of a single chain of N monomers

$$S(k,t) = \frac{1}{N} \left\langle \sum_{ij=1}^{N} \exp\left[i\mathbf{k} \cdot (\mathbf{r}_i(t) - \mathbf{r}_j(0))\right] \right\rangle \tag{311}$$

scales as

$$S(k,t) = k^{-1/\nu} f\left(k^2 t^{2/z}\right). \tag{312}$$

The Zimm model applies to dilute solutions, and, therefore, to the dynamics of a single solvated chain. It has become a benchmark system, used to test the validity of mesoscopic simulation methods. A single chain, modeled by bead–spring interactions, coupled to a surrounding solvent to account for hydrodynamic interactions, has been successfully simulated via (1) Molecular Dynamics [180–182], (2) Dissipative Particle Dynamics [183, 184], Multi-Particle Collision Dynamics [185, 186], and by LB [122], applying the dissipative coupling described in Sect. 4.5. These studies are nowadays all sufficiently accurate to be able to clearly distinguish between Rouse and Zimm scaling. However, a more demanding goal is to verify not only the exponent, but also the prefactor of the dynamic scaling law.

The Kirkwood approximation to the diffusion constant,

$$D^{(K)} = \frac{k_B T}{\Gamma N} + \frac{k_B T}{6\pi\eta} \left\langle \frac{1}{R_H} \right\rangle, \tag{313}$$

(Γ is the monomer friction coefficient) can be calculated from a conformational average of the hydrodynamic radius R_H,

$$\left\langle \frac{1}{R_H} \right\rangle = \frac{1}{N^2} \sum_{i \neq j} \left\langle \frac{1}{r_{ij}} \right\rangle. \tag{314}$$

Highly accurate results for a single chain in a structureless solvent have been obtained by Monte Carlo methods [187, 188]. However, a naive comparison will fail badly. The expression for R_H (314) assumes an infinite system, but in a simulation the system is confined to a periodic unit cell, which is typically not substantially larger than the size of the coil. The effects of periodic boundaries can be accounted for quantitatively, by replacing the Oseen tensor with an Ewald sum that includes the hydrodynamic interactions with the periodic images [167]. The consequences of this have been worked out in detail for polymers [182] and colloids [19]. The main result is that the hydrodynamic radius must be replaced by a system-size dependent effective hydrodynamic radius; the leading-order correction is proportional to R/L, where L is the linear dimension of the periodic simulation cell. Interestingly, *internal* modes, such as Rouse modes [6], where the motion of the center of mass has been subtracted, have a much weaker finite-size effect, which scales as L^{-3}, corresponding to a dipolar hydrodynamic interaction with the periodic images [122]. Taking the finite-size effects into account, the predictions of the Zimm model are nicely confirmed.

However, the Zimm model is no longer valid as soon as the chains start to overlap. Here a double screening mechanism sets in: (1) Screening of excluded volume interactions (Flory screening). In a dense melt, the chain conformations are not those of a self-avoiding walk, but rather those of a random walk ($\nu = 1/2$). Essentially, this is an entropic packing effect: A swollen coil would take too much configuration space from the surrounding chains. This effect can be understood in terms of a self-consistent mean-field theory, which is expected to work well for dense systems where density fluctuations are suppressed [7]. (2) Screening of hydrodynamic interactions. In dense melts, the dynamics is not Zimm-like, but rather Rouse-like, or governed by reptation [6]. Reptation occurs for long chains in dense systems, where topological constraints enforce an essentially curvilinear motion. We will not be concerned with these latter effects, but rather with the mechanism which leads to the suppression of hydrodynamic interactions. On the basis of the results of computer simulations [43], we were able to develop a simple picture, which essentially confirmed the previous work by de Gennes [189], and completed it. The basic mechanism is chain–chain collisions. A monomer encountering another chain will deform it elastically, inducing a stress along the polymer backbone instead of propagating the signal into the surroundings. Since the chain arrangements are random, the fluid momentum is also randomized, such that momentum correlations (or hydrodynamic interactions) are destroyed.

The crossover region between dilute and dense systems is called the semidilute regime. A semidilute solution is characterized by strongly overlapping chains which are however so long and so dilute that the monomer concentration can still be considered as vanishingly small. Apart from b and R, there is now a third important length scale, the "blob size" ξ with $b \ll \xi \ll R$. Essentially, ξ is the length scale on which interactions with the surrounding chains become important; this length scale controls the crossover from dilute to dense behavior. The chain conformations are characterized by $\nu = 0.59$ on length scales much smaller than ξ, while on length scales substantially above ξ the exponent is $\nu = 1/2$. The challenge for computer simulations is that both behaviors need to be resolved simultaneously, which is only possible for $N > 10^3$. Roughly 30 monomers are needed to resolve the random fractal structure within the blob, while another 30 blobs per chain are needed to observe the random walk regime. Furthermore, a many-chain system should be run without self-overlaps, and this leads to the conclusion [43] that the smallest system to simulate semidilute dynamics contains roughly 5×10^4 monomers and 5×10^5 LB lattice sites.

The picture which emerges from these simulations [43] can be summarized as follows. Initially, the dynamics is Zimm-like, even for length scales beyond the blob size. The reason is that hydrodynamic signals can spread easily throughout the system, and just drag the chains with them. This continues until chain–chain collisions start to play a role. The relevant time scale is the blob relaxation time $\tau_\xi \propto \xi^3$, i.e., the time a blob needs to move its own size. From then on, the screening mechanism described above becomes important, and the dynamics is Rouse-like. This is only observable on length scales beyond the blob size, since on smaller scales all dynamic correlations have decayed already. This is nicely borne out by single-chain dynamic structure factor data (see Fig. 10), and explains the previous observation of "incomplete screening" [190] in a straightforward and natural way.

5.3 Polymer Migration in Confined Geometries

Flexible polymers in a pressure-driven flow field migrate towards the center of the channel, because of hydrodynamic interactions. The local shear rate stretches the polymer and the resulting tension in the chain generates an additional flow field around the polymer. This flow field becomes asymmetric near a no-slip boundary and results in a net drift towards the center of the channel [191–193]. Recent simulations [193, 194] show that hydrodynamic lift is the dominant migration mechanism in pressure-driven flow, rather than spatial gradients in shear rate. Recently, we used

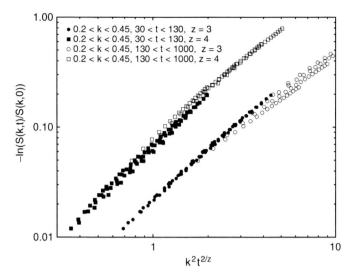

Fig. 10 Scaling of the single-chain dynamic structure factor data, showing both Rouse and Zimm scaling [43]. The wave number k has been restricted such that only length scales above the blob size are probed ($k\xi < 1$), while the size of the polymer chain as a whole does not yet matter ($kR_G > 1$). The data are labeled according to the time regimes; *solid symbols* refer to the short-time regime below the blob relaxation time, $t < \tau_\xi$, while *open symbols* are for later times $t > \tau_\xi$. The *upper curve* is for Rouse scaling ($z = 4$) and the *lower curve* for Zimm scaling ($z = 3$). One sees that Zimm scaling works better in the short-time regime, while Rouse scaling holds for later times

numerical simulations to investigate a flexible polymer driven by a combination of fluid flow and external body force [195], but ignoring the complications arising from counterion screening in electrophoretic flows. We used the fluctuating LB model (Sect. 3) in conjunction with the point-force coupling scheme described in Sect. 4.5.

We were surprised to find that the polymer migrates towards the channel center under the action of a body-force alone, while in combination with a pressure-driven flow the polymer can move either towards the channel wall or towards the channel center. The external field and pressure gradient result in two different Peclet numbers: $Pe = \overline{U}R_g/D$ and $Pe_f = \overline{\gamma}R_g{}^2/D$. Here $\overline{U}$ is the average polymer velocity with respect to the fluid, and $\overline{\gamma}$ is the average shear rate. The interplay between force and flow can lead to a wide variety of steady-state distributions of the polymer center of mass across the channel [195]. For example, in a countercurrent application of the two fields, the polymer tends to orient in different quadrants depending on the relative magnitude of the two driving forces. The polymer then drifts either towards the walls or towards the center depending on its mean orientation. The results in Fig. 11 show migration towards the boundaries when the external force is small ($Pe < 30$), but increasing the force eventually reverses the orientation of the polymer and the polymer again migrates towards the center ($Pe > 100$).

The simulations mimic recent experimental observations of the migration of DNA in combined electric and pressure-driven flow fields [196,197]. The similarities

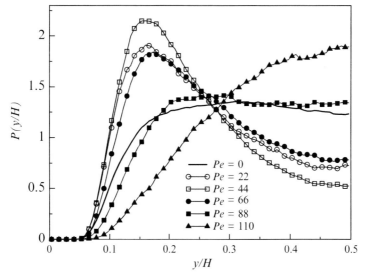

Fig. 11 Center of mass distributions for countercurrent application of an external body force and pressure-driven flow in a channel of width H. The *solid curve* shows the level of migration under the pressure-driven flow only. The flow Peclet number in all cases is $Pe_f = 12.5$. The boundary is at $y/H = 0$ and the center of the channel is at $y/H = 0.5$; $H = 8R_g$

between these results suggest that hydrodynamic interactions in polyelectrolyte solutions are only partially screened. In fact, within the Debye–Hückel approximation, there is a residual dipolar flow field [198]. Although this flow is weak in comparison to the electrophoretic velocity, its dipolar orientation enables it to drive a transverse migration of the polymer. A recent kinetic theory calculation [199] supports these qualitative observations.

5.4 Sedimentation

The previous examples have focused on sub-micrometer-sized particles, colloids and polymers, where Brownian motion is an essential component of the dynamics. For particles larger than a few micrometers, Brownian motion is negligible under normal laboratory conditions and a suspension of such particles can be simulated using the deterministic version of the LB equation (see Sect. 3). There is an interesting regime of particle sizes, from 1–$100\,\mu\text{m}$ depending on solvent, where Brownian motion is negligible, yet inertial effects are still unimportant. This corresponds to the region of low Reynolds number ($Re = Ud/\eta_{\text{kin}}$) but high Peclet number ($Pe = Ud/D$). Here U is the characteristic particle velocity, d is the diameter, and D is the particle diffusion coefficient. Because of the large difference in time scale between diffusion of momentum and particle diffusion, it is quite feasible for

Pe to be 6–10 orders of magnitude larger than *Re*. We have carried out a number of simulations in this regime, with the aim of elucidating the role of suspension microstructure in controlling the amplitude of the velocity fluctuations as the suspension settles.

In a sedimenting suspension, spatial and temporal variations in particle concentration drive large fluctuations in the particle velocities, of the same order of magnitude as the mean settling velocity. For particles larger than a few micrometers, this hydrodynamic diffusion dominates the thermal Brownian motion, and in the absence of inertia ($Re \ll 1$), the particle velocities are determined entirely by the instantaneous particle positions. If the particles are randomly distributed, then the velocity fluctuations will diverge with increasing container size [200], although the density fluctuations may eventually drain out of the system by convection [201]. However, experimental measurements indicate that the velocity fluctuations converge to a finite value as the container dimensions are increased [202, 203], but the mechanism by which the velocity fluctuations saturate is not yet clear. Some time ago, Koch and Shaqfeh suggested that the distribution of pairs of particles could be modified by shearing forces induced by the motion of a third particle, and that these changes in microstructure could in turn lead to a screening of the long-range hydrodynamic interactions driving the velocity fluctuations [204]. However, detailed numerical simulations found no evidence of the predicted microstructural changes [205]. Instead the velocity fluctuations in homogeneous suspensions (with periodic boundary conditions) were found to diverge with increasing cell size. More recently, it has been proposed that long wavelength density fluctuations can be suppressed by a convection diffusion mechanism [206, 207], but a bulk screening mechanism cannot be reconciled with the results of computer simulations [144,205]. Alternatively, it has been suggested that the vertical walls of the container may modify, although not eliminate, the divergence of the velocity fluctuations [208]. Most recently, it has been shown [209,210] that a small vertical density gradient can damp out diverging velocity fluctuations.

LB simulations were used to test these theoretical ideas, comparing the behavior of the velocity fluctuations in three different geometries [211]. We found striking differences in the level of velocity fluctuations, depending on the macroscopic boundary conditions. In a geometry similar to those used in laboratory experiments [202, 203], namely a rigid container bounded in all three directions, we found that the calculated velocity fluctuations saturate with increasing container dimensions, as observed experimentally, but contrary to earlier simulations with periodic boundary conditions [144, 205]. The main result is illustrated in Fig. 12, and suggests that the velocity fluctuations in a bounded container are independent of container width for sufficiently large containers. On the other hand, in vertically homogeneous suspensions velocity fluctuations are proportional to the container width, regardless of the boundary conditions in the horizontal plane. The significance of this result is that it establishes that velocity fluctuations in a sedimenting suspension depend on the macroscopic boundary conditions and that laboratory measurements [202,203,212] are not necessarily characteristic of a uniform suspension, as had been supposed. Instead, the simulations show that vertical

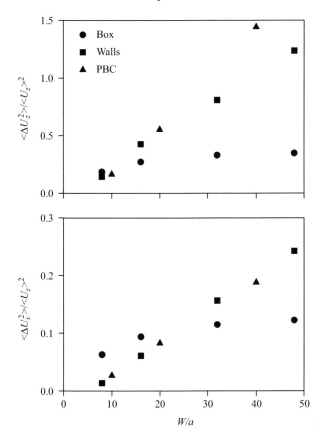

Fig. 12 Relative velocity fluctuations $\langle \Delta U_\alpha^2 \rangle / \langle U_z \rangle^2$ as a function of container width. The vertical, $\langle \Delta U_z^2 \rangle$, and horizontal, $\langle \Delta U_x^2 \rangle$, fluctuations are shown for three different boundary conditions: a cell bounded in all three directions by no-slip walls (Box), a cell bounded by vertical walls (Walls), and a cell that is periodic in all three directions (PBC) [144, 205]. The statistical errors are comparable to the size of the symbols

variations in particle concentration are responsible for suppressing the velocity fluctuations, which otherwise diverge with increasing container size, in agreement with theory [200] and earlier simulations [144, 205]. The upper and lower boundaries apparently act as sinks for the fluctuation energy [201], while in homogeneous suspensions velocity fluctuations remain proportional to the system size [200].

5.5 Inertial Migration in Pressure-Driven Flow

At still larger particle sizes, typically in excess of 100 μm, the inertia of the fluid can no longer be ignored. For suspended particles in a gravitational field, the Reynolds number grows in proportion to the cube of the particle size. Inertia breaks the symmetry inherent in Stokes flow and leads to new phenomena, and in particular the possibility of lateral migration of particles. A particle in a shear flow experiences a transverse force at non-zero Re, with a direction that depends on the velocity of the particle with respect to the fluid velocity at its center. Thus, if the particle is moving slightly faster than the fluid it moves crosswise to the flow in the direction of lower fluid velocity and vice versa [213]; if it is moving with the local stream velocity then it does not migrate in the lateral direction at all. Now in Poiseuille flow, a spherical particle moves faster than the surrounding fluid because of the Faxen force, proportional to the curvature in the fluid velocity field. Thus, particles tend to migrate towards the channel walls [214]. However, near the wall the particle is slowed down by the additional drag with the wall and so eventually migrates the other way. At small Reynolds numbers ($Re < 100$), these forces balance when the particle is at a radial position of roughly $0.6R$, where R is the radius of the pipe. In a cylindrical pipe, a uniformly distributed suspension of particles rearranges to form a stable ring located at approximately $0.6R$ [215]. Theoretical calculations for small particles in plane Poiseuille flow give similar equilibrium positions to those observed experimentally [216, 217]. The profile of the lateral force across the channel shows only one equilibrium position, which shifts closer to the boundary wall as the Reynolds number increases. Our interest in this problem was sparked by two recent experimental observations: first that particles tend to align near the walls to make linear chains of more or less equally spaced particles [215, 218], and second that at high Reynolds numbers ($Re \sim 1,000$) an additional inner ring of particles was observed when the ratio of particle diameter d to cylinder diameter D was of the order of 1:10 [219]. Large particles introduce an additional Reynolds number, $Re_p = Re(d/D)^2$, which may not be small, as assumed theoretically [216, 217]. We used the LB method to investigate inertial migration of neutrally buoyant particles in the range of Reynolds numbers from 100 to 1,000 [220]. Individual particles in a channel with a square cross section migrate to one of a small number of equilibrium positions in the cross-sectional plane, located near an edge or at the center of a face; we could not identify any stable positions for single particles near the center of the channel.

To investigate multiparticle suspensions, random configurations of particles were prepared at a volume fraction $\phi = 1\%$ and size ratio $H/d = 9.1$. The Reynolds number in the simulations varied between 100 and 1,000. An initially uniform distribution, shown in Fig. 13a, evolves into three different steady-state distributions depending on Re. At $Re = 100$ (Fig. 13b) particles are gathered around the eight equilibrium positions and strongly aligned in the direction of the flow, making linear chains of more or less uniformly spaced particles. Similar trains of particles were observed in laboratory experiments [218]. At $Re = 500$ (Fig. 13c) the particles are gathered in one of the four most stable positions, near each corner. By a Reynolds number of 500, the trains are unstable and the spacing between the particles is no longer uniform. Instead transient aggregates of closely spaced particles are formed, again near the corners of the duct. However, at still higher Reynolds number, $Re = 1,000$, there is another change in particle configuration (Fig. 13d),

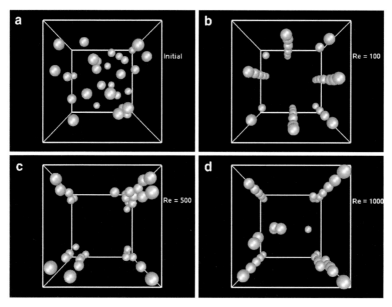

Fig. 13 Snapshots of particle configurations in a duct flow at different Reynolds numbers; the flow is into the plane of the paper: **a** Initial configuration, **b** $Re = 100$, **c** $Re = 500$, **d** $Re = 1,000$. The ratio $H/d = 9.1$, the number of particles $N = 32$ and the volume fraction $\phi = 1\%$

and particles appear in the center of the duct. A central band was first observed in experiments in a cylindrical pipe [219], but its origin remains unclear. We observe that the central particles have a substantial diffusive motion in the velocity-gradient plane, whereas the particle trains exhibit little transverse diffusion. Since there are no single-particle equilibrium positions at the duct center, the presence of particles in the inner region is clearly due to multi-body interactions. Nevertheless, this migration cannot be a shear-induced migration of the kind that occurs in low-Reynolds number flows [221].

Our simulations suggest that the inner band of particles is the result of the formation of transient clusters of particles. We proposed [220] that at higher Reynolds numbers ($Re > 500$) the trains become unstable and clusters of closely spaced particles arise, as can be seen in Fig. 13c. Simulations of tethered pairs of particles have shown that additional equilibrium positions arise for pairs of particles at Reynolds numbers in excess of 750 [220]. Thus, transient clusters are formed at higher Reynolds numbers, which drift towards the center of the channel making the additional ring observed in experiments [219] and simulations (Fig. 13d). Eventually, the cluster disintegrates from hydrodynamic dispersion and the particles return to the walls. At steady state, there is a flux of pairs and triplets of particles moving towards the center, balanced by individual particles moving towards the walls. This also explains why the particles in the inner region are highly mobile, while those near the walls have a very small diffusivity.

Acknowledgements We thank E. Wajnryb – IPPT (Institute of Fundamental Research), Warsaw – for access to the multipole code used in Sect. 4. A.J.C. Ladd thanks the Alexander von Humboldt Foundation for supporting his stay at the Max Planck Institute for Polymer Research by a Humboldt Research Award.

References

1. Chaikin PM, Lubensky TC (1997) Principles of condensed matter physics. Cambridge University Press, Cambridge
2. Russel WB, Saville DA, Schowalter WR (1995) Colloidal dispersions. Cambridge University Press, Cambridge
3. Lyklema J (1991) Fundamentals of interface and colloid science. Academic, London
4. Evans DF, Wennerström H (1999) The colloidal domain, 2nd edn. Wiley, New York
5. de Gennes PG (1979) Scaling concepts in polymer physics. Cornell University Press, Ithaca
6. Doi M, Edwards SF (1986) The theory of polymer dynamics. Oxford University Press, Oxford
7. Grosberg AY, Khokhlov AR (1994) Statistical physics of macromolecules. AIP, New York
8. Nelson P (2007) Biological physics. W. H. Freeman, New York
9. Alder BJ, Wainwright TE (1970) Phys Rev A 1:18
10. Montfrooij W, de Schepper I (1989) Phys Rev A 39:2731
11. Bird RB, Curtiss CF, Armstrong RC, Hassager O (1987) Dynamics of polymeric liquids. Wiley, New York
12. Grest GS, Kremer K (1986) Phys Rev A 33:3628
13. Weeks JD, Chandler D, Andersen HC (1971) J Chem Phys 54:5237
14. Hansen JP, McDonald IR (1986) Theory of simple liquids. Academic, London
15. Risken H (1984) The Fokker–Planck equation. Springer, Berlin
16. Gardiner CW (1985) Handbook of stochastic methods for physics, chemistry, and the natural sciences. Springer, Berlin
17. Mazur P, Saarloos WV (1982) Physica A 115:21
18. Cichocki B, Felderhof BU (1988) J Chem Phys 89:3705
19. Ladd AJC (1990) J Chem Phys 93:3484
20. Sangani AS, Mo G (1996) Phys Fluids 8:1990
21. Cichocki B, Jones RB, Kutteh R, Wajnryb E (2000) J Chem Phys 112:2548
22. Happel J, Brenner H (1986) Low-Reynolds number hydrodynamics. Martinus Nijhoff, Dordrecht
23. Rotne J, Prager S (1969) J Chem Phys 50:4831
24. Wajnryb E, Szymczak P, Cichocki B (2004) Physica A 335:339
25. Öttinger HC (1996) Stochastic processes in polymeric fluids. Springer, Berlin
26. Kröger M (2004) Phys Rep 390:453
27. Ermak DL, McCammon JA (1978) J Chem Phys 69:1352
28. Liu B, Dünweg B (2003) J Chem Phys 118:8061
29. Brady JF, Bossis G (1988) Ann Rev Fluid Mech 20:111
30. Fixman M (1986) Macromolecules 19:1204
31. Jendrejack RM, Graham MD, dePablo JJ (2000) J Chem Phys 113:2894
32. Sierou A, Brady JF (2001) J Fluid Mech 448:115
33. Banchio AJ, Brady JF (2003) J Chem Phys 118:10323
34. Saintillan D, Darve E, Shaqfeh ESG (2005) Macromolecules 17:33301
35. Hernandez-Ortiz JP, de Pablo JJ, Graham MD (2007) Phys Rev Lett 98:140602
36. Meng Q, Higdon JJL (2008) J Rheol 52:1
37. Meng Q, Higdon JJL (2008) J Rheol 52:37
38. Blake JR (1971) Proc Camb Philos Soc 70:303
39. Liron N, Mochon S (1976) J Eng Math 10:287
40. Cichocki B, Jones RB (1998) Physica A 258:273

41. Jendrejack RM, Graham MD, dePablo JJ (2003) J Chem Phys 119:1165
42. Landau LD, Lifshitz EM (1959) Fluid mechanics. Addison-Wesley, London
43. Ahlrichs P, Everaers R, Dünweg B (2001) Phys Rev E 64:040501 (R)
44. Hoogerbrugge PJ, Koelman JMVA (1992) Europhys Lett 19:155
45. Koelman JMVA, Hoogerbrugge PJ (1993) Europhys Lett 21:369
46. Español P, Warren P (1995) Europhys Lett 30:191
47. Español P (1995) Phys Rev E 52:1734
48. Groot R, Warren P (1997) J Chem Phys 107:4423
49. Español P (1998) Phys Rev E 57:2930
50. Pagonabarraga I, Hagen MJH, Frenkel D (1998) Europhys Lett 42:377
51. Gibson JB, Chen K, Chynoweth S (1999) Int J Mod Phys C 10:241
52. Besold G, Vattulainen I, Karttunen M, Polson JM (2000) Phys Rev E 62:R7611
53. Vattulainen I, Karttunen M, Besold G, Polson JM (2002) J Chem Phys 116:3967
54. Nikunen P, Karttunen M, Vattulainen I (2003) Comput Phys Commun 153:407
55. Shardlow T (2003) SIAM J Sci Comp 24:1267
56. Soddemann T, Dünweg B, Kremer K (2003) Phys Rev E 68:046702
57. Junghans C, Praprotnik M, Kremer K (2008) Soft Matter 4:156
58. Malevanets A, Kapral R (1999) J Chem Phys 110:8605
59. Ihle T, Kroll DM (2003) Phys Rev E 67:066705
60. Ihle T, Kroll DM (2003) Phys Rev E 67:066706
61. Kikuchi N, Pooley CM, Ryder JF, Yeomans JM (2003) J Chem Phys 119:6388
62. Ihle T, Tüzel E, Kroll DM (2004) Phys Rev E 70:035701(R)
63. Sharma N, Patankar NA (2004) J Comput Phys 201:466
64. Car R, Parrinello M (1985) Phys Rev Lett 55:2471
65. Höfler K, Schwarzer S (2000) Phys Rev E 61:7146
66. Schwarzer S (1995) Phys Rev E 52:6461
67. Kalthoff W, Schwarzer S, Herrmann HJ (1997) Phys Rev E 56:2234
68. Wachmann B, Kalthoff W, Schwarzer S, Herrmann HJ (1998) Granular Matter 1:75
69. Delgado-Buscalioni R, Coveney PV (2003) Phys Rev E 67:046704
70. Delgado-Buscalioni R, Flekkoy EG, Coveney PV (2005) Europhys Lett 69:959
71. Delgado-Buscalioni R, Coveney PV, Riley GD, Ford RW (2005) Philos Trans Math Phys Eng Sci 363:1975
72. Fabritiis GD, Serrano M, Delgado-Buscalioni R, Coveney PV (2007) Phys Rev E 75:026307
73. Giupponi G, Fabritiis GD, Coveney PV (2007) J Chem Phys 126:154903
74. Delgado-Buscalioni R, Fabritiis GD (2007) Phys Rev E 76:036709
75. Frisch U, Hasslacher B, Pomeau Y (1986) Phys Rev Lett 56:1505
76. Frisch U, d'Humières D, Hasslacher B, Lallemand P, Pomeau Y, Rivet JP (1987) Complex Syst 1:649
77. Succi S (2001) The lattice Boltzmann equation for fluid dynamics and beyond. Oxford University Press, Oxford
78. McNamara GR, Zanetti G (1988) Phys Rev Lett 61:2332
79. Higuera F, Succi S, Benzi R (1989) Europhys Lett 9:345
80. Benzi R, Succi S, Vergassola M (1992) Phys Rep 222:145
81. Alexander FJ, Chen S, Sterling JD (1993) Phys Rev E 47:R2249
82. Ihle T, Kroll D (2000) Comput Phys Commun 129:1
83. Lallemand P, Luo LS (2003) Phys Rev E 68:036706
84. Guo Z, Zheng C, Shi B, Zhao TS (2007) Phys Rev E 75:036704
85. Ansumali S, Karlin IV (2005) Phys Rev Lett 95:260605
86. Prasianakis NI, Karlin IV (2007) Phys Rev E 76:016702
87. Gunstensen AK, Rothman DH, Zaleski S, Zanetti G (1991) Phys Rev A 43:4320
88. Shan X, Chen H (1993) Phys Rev E 47:1815
89. Swift MR, Orlandini E, Osborn WR, Yeomans JM (1996) Phys Rev E 54:5041
90. Gonnella G, Orlandini E, Yeomans JM (1999) Phys Rev E 59:R4741
91. Luo LS (2000) Phys Rev E 62:4982
92. Luo LS, Girimaji SS (2003) Phys Rev E 67:036302

93. Guo Z, Zhao TS (2005) Phys Rev E 71:026701
94. Arcidiacono S, Mantzaras J, Ansumali S, Karlin IV, Frouzakis C, Boulouchos KB (2006) Phys Rev E 74:056707
95. Halliday I, Hollis AP, Care CM (2007) Phys Rev E 76:026708
96. Li Q, Wagner AJ (2007) Phys Rev E 76:036701
97. Qian YH, D'Humieres D, Lallemand P (1992) Europhys Lett 17:479
98. Ladd AJC (1994) J Fluid Mech 271:285
99. Ladd AJC (1994) J Fluid Mech 271:311
100. Ladd AJC, Verberg R (2001) J Stat Phys 104:1191
101. Adhikari R, Stratford K, Cates ME, Wagner AJ (2005) Europhys Lett 3:473
102. Dünweg B, Schiller UD, Ladd AJC (2007) Phys Rev E 76:036704
103. Hinch EJ (1991) Perturbation methods. Cambridge University Press, Cambridge
104. Nguyen NQ, Ladd AJC (2002) Phys Rev E 66:046708
105. Chapman S, Cowling TG (1960) The mathematical theory of non-uniform gases. Cambridge University Press, Cambridge
106. McNamara GR, Alder BJ (1992) In: Mareschal M, Holian BL (eds) Microscopic simulations of complex hydrodynamic phenomena. Plenum, New York
107. Wagner AJ (1998) Europhys Lett 44:144
108. Karlin IV, Ferrante A, Öttinger HC (1999) Europhys Lett 47:182
109. Boghosian BM, Love PJ, Coveney PV, Karlin IV, Succi S, Yepez J (2003) Phys Rev E 68:025103(R)
110. D'Humières D, Ginzburg I Krafczyk M, Lallemand P, Luo LS (2002) Philos Trans R Soc Lond A 360:437
111. Lallemand P, Luo LS (2000) Phys Rev E 61:6546
112. D'Humières D (1992) Prog Astronaut Aeronaut 159:450
113. Ginzburg I, d'Humières D (2003) Phys Rev E 68:066614
114. Chun B, Ladd AJC (2007) Phys Rev E 75:066705
115. McNamara GR, Alder BJ (1993) Physica A 194:218
116. Landau DP, Binder K (2000) A guide to Monte Carlo simulations in statistical physics. Cambridge University Press, Cambridge
117. Ginzburg I, Adler PM (1994) J Phys II France 4:191
118. Guo Z, Zheng C, B Shi B (2002) Phys Rev E 65:046308
119. Ladd AJC, Frenkel D (1989) In: Manneville P, Boccara N, Vichniac GY, Bidaux R (eds) Cellular automata and modeling of complex physical systems. Springer Proceedings in Physics no. 46. Springer, Berlin, pp. 242–245
120. Ladd AJC, Frenkel D (1990) Phys Fluids A 2:1921
121. Ahlrichs P, Dünweg B (1998) Int J Mod Phys C 9:1429
122. Ahlrichs P, Dünweg B (1999) J Chem Phys 111:8225
123. Fyta MG, Melchionna S, Kaxiras E, Succi S (2006) Multiscale Model Simul 5:1156
124. Lobaskin V, Dünweg B (2004) New J Phys 6:54
125. Lobaskin V, Dünweg B, Holm C (2004) J Phys Cond Matt 16:S4063
126. Lobaskin V, Dünweg B, Medebach M, Palberg T, Holm C (2007) Phys Rev Lett 98:176105
127. Chatterji A, Horbach J (2005) J Chem Phys 122:184903
128. Chatterji A, Horbach J (2007) J Chem Phys 126:064907
129. Peskin CS (2002) Acta Numer 11:479
130. Nash RW, Adhikari R, Cates ME (2008) Phys Rev E 77:026709
131. Feng ZG, Michaelides E (2003) J Comput Phys 195:602
132. Shi X, Phan-Thien N (2005) J Comput Phys 206:81
133. Aidun CK, Lu YN, Ding E (1998) J Fluid Mech 373:287
134. Lowe CP, Frenkel D, Masters AJ (1995) J Chem Phys 103:1582
135. Ricci A, Ciccotti G (2003) Mol Phys 101:1927
136. Bussi G, Parrinello M (2007) Phys Rev E 75:056707
137. Thalmann F, Farago J (2007) J Chem Phys 127:124109
138. Fabritiis GD, Serrano M, Español P, Coveney P (2006) Physica A 361:429
139. Serrano M, Fabritiis GD, Español P, Coveney P (2006) Math Comput Simul 72:190

140. Usta OB, Ladd AJC, Butler JE (2005) J Chem Phys 122:094902
141. Ghadder CK (1995) Phys Fluids 7:2563
142. Ding EJ, Aidun CK (2003) J Stat Phys 112:685
143. Claeys IL, Brady JF (1989) PhysicoChem Hydrodyn II:261
144. Ladd AJC (1997) Phys Fluids 9:491
145. Brady JF, Durlofsky LJ (1988) Phys Fluids 31:717
146. Buxton GA, Verberg R, Jasnow D, Balazs AC (2005) Phys Rev E 71:056707
147. Alexeev A, Verberg R, Balazs AC (2006) Soft Matter 2:499
148. Junk M, Yang Z (2005) J Stat Phys 121:3
149. Chen HD (1998) Phys Rev E 58:3955
150. Chen HD, Teixeira C, Molvig K (1998) Int J Mod Phys C 9:1281
151. Bouzidi M, Firdaouss M, Lallemand P (2001) Phys Fluids 13:3452
152. Filippova O, Hänel D (1998) J Comput Phys 147:219
153. Mei RW, Luo LS, Shyy W (1999) J Comput Phys 155:307
154. Lallemand P, Luo LS (2003) J Comput Phys 184:406
155. Maxey MR, Patel BK (2001) Int J Multiphase Flow 27:1603
156. Lomholt S, Stenum B, Maxey MR (2002) Int J Multiphase Flow 28:225
157. Zakharov VE, Kuznetsov EA (1997) Phys Usp 40:1087
158. Chandrasekhar S (1943) Rev Mod Phys 15:1
159. Dünweg B (2003) In: Dünweg B, Landau DP, Milchev AI (eds) Computer simulations of surfaces and interfaces. Kluwer, Dordrecht
160. Fox RF, Uhlenbeck GE (1970) Phys Fluids 18:1893
161. http://en.wikipedia.org
162. van Gunsteren W, Berendsen HJC (1988) Molec Simul 1:173
163. McLachlan RI, Quispel GRW (2002) Acta Numer 11:341
164. Forbert HA, Chin SA (2000) Phys Rev E 63:016703
165. Hasimoto H (1959) J Fluid Mech 5:317
166. Ladd AJC (2000) In: van Beijeren H, Karkheck J (eds) Dynamics: models and kinetic methods for non-equilibrium many body systems. Kluwer, Dordrecht, pp. 17–30
167. Beenakker CWJ (1986) J Chem Phys 85:1581
168. Ladd AJC (1988) J Chem Phys 88:5051
169. Weitz DA, Pine DJ, Pusey PN, Tough RJA (1989) Phys Rev Lett 63:1747
170. Ernst MH, Hauge EH, van Leeuwen JMJ (1970) Phys Rev Lett 25:1254
171. Dorfman JR, Cohen EGD (1975) Phys Rev A 12:292
172. van der Hoef MA, Frenkel D, Ladd AJC (1991) Phys Rev Lett 67:3459
173. van der Hoef MA, Frenkel D (1991) Phys Rev Lett 66:1591
174. Ladd AJC (1993) Phys Rev Lett 70:1339
175. Hauge EH, Martin-Löf A (1973) J Stat Phys 7:259
176. Zhu JX, Durian DJ, Müller J, Weitz DA, Pine DJ (1992) Phys Rev Lett 68:2559
177. Kao MH, Yodh AG, Pine DJ (1993) Phys Rev Lett 70:242
178. Zimm BH (1956) J Chem Phys 24:269
179. Rouse PE (1953) J Chem Phys 21:1272
180. Pierleoni C, Ryckaert JP (1992) J Chem Phys 96:8539
181. Smith W, Rapaport DC (1992) Mol Sim 9:25
182. Dünweg B, Kremer K (1993) J Chem Phys 99:6983
183. Schlijper AG, Hoogerbrugge PJ, Manke CW (1995) J Rheol 39:567
184. Spenley NA (2000) Europhys Lett 49:534
185. Malevanets A, Yeomans JM (2000) Europhys Lett 52:231
186. Mussawisade K, Ripoll M, Winkler RG, Gompper G (2005) J Chem Phys 123:144905
187. Ladd AJC, Frenkel D (1992) Macromolecules 25:3435
188. Dünweg B, Reith D, Steinhauser M, Kremer K (2002) J Chem Phys 117:914
189. de Gennes PG (1976) Macromolecules 9:594
190. Richter D, Binder K, Ewen B, Stühn B (1984) J Phys Chem 88:6618
191. Fang L, Hu H, Larson RG (2005) J Rheol 49:127
192. Jendrejack RM, Schwartz DC, de Pablo JJ, Graham MD (2004) J Chem Phys 120:2513

193. Ma H, Graham M (2005) Phys Fluids 17:083103
194. Usta OB, Butler JE, Ladd AJC (2006) Phys Fluids 18:031703
195. Usta OB, Butler JE, Ladd AJC (2007) Phys Rev Lett 98:090831
196. Zheng J, Yeung ES (2002) Anal Chem 74:4536
197. Zheng J, Yeung ES (2003) Anal Chem 75:3675
198. Long D, Ajdari A (2001) Euro Phys J E 4:29
199. Butler JE, Usta OB, Kekre R, Ladd AJC (2007) Phys Fluids 19:113101
200. Caflisch RE, Luke JHC (1985) Phys Fluids 28:759
201. Hinch EJ (1988) In: Guyon E, Pomeau Y, Nadal JP (eds) Disorder and mixing. Kluwer, Dordrecht, pp. 153–161
202. Nicolai H, Guazzelli E (1995) Phys Fluids 7:3
203. Segré PN, Herbolzheimer E, Chaikin PM (1997) Phys Rev Lett 79:2574
204. Koch DL, Shaqfeh ESG (1991) J Fluid Mech 224:275
205. Ladd AJC (1996) Phys Rev Lett 76:1392
206. Tong P, Ackerson BJ (1998) Phys Rev E 58:R6931
207. Levine A, Ramaswamy S, Frey E, Bruinsma R (1998) Phys Rev Lett 81:5944
208. Brenner MP (1999) Phys Fluids 11:754
209. Luke JHC (2000) Phys Fluids 12:1619
210. Mucha PJ, Brenner MP (2003) Phys Fluids 15:1305
211. Ladd AJC (2002) Phys Rev Lett 88:048301
212. Guazzelli E (2001) Phys Fluids 13:1537
213. Saffman PG (1965) J Fluid Mech 22:385
214. Ho BP, Leal LG (1974) J Fluid Mech 65:365
215. Segré G, Silberberg A (1962) J Fluid Mech 14:136
216. Schonberg JA, Hinch EJ (1989) J Fluid Mech 203:517
217. Asmolov ES (1999) J Fluid Mech 381:63
218. Matas J, Glezer V, Guazzelli E, Morris J (2004) Phys Fluids 16:4192
219. Matas JP, Morris JF, Guazzelli E (2004) J Fluid Mech 515:171
220. Chun B, Ladd AJC (2006) Phys Fluids 18:031704
221. Frank M, Anderson D, Weeks ER, Morris JF (2003) J Fluid Mech 493:363

Adv Polym Sci 221: 167–233
DOI:10.1007/12_2008_3

Transition Path Sampling and Other Advanced Simulation Techniques for Rare Events

Christoph Dellago and Peter G. Bolhuis

Abstract Computer simulations of molecular processes such as nucleation in first-order phase transitions or the folding of a protein are often complicated by widely disparate time scales related to important but rare events. Here, we will review several recently developed computational methods designed to address the rare-events problem. In doing so, we will focus on the transition path sampling methodology.

Keywords Computer simulation, Rare events, Rate constant, Reaction mechanism, Transition pathway

Contents

C. Dellago (✉)
Faculty of Physics, University of Vienna, Boltzmanngasse 5, 1090 Vienna, Austria
e-mail: Christoph.Dellago@univie.ac.at

P.G. Bolhuis
van't Hoff Institute for Molecular Sciences, University of Amsterdam, Nieuwe Achtergracht 166, 1018 WV Amsterdam, The Netherlands
e-mail: bolhuis@science.uva.nl

List of Abbreviations and Symbols

CMD	Coarse molecular dynamics
DPS	Discrete path sampling
FES	Free energy surface
FFS	Forward flux sampling
GNN	Genetic neural network
KMC	Kinetic Monte Carlo
MC	Monte Carlo
MD	Molecular dynamics
MEP	Minimum energy path
MFEP	Minimum free energy path
MLE	Maximum likelihood estimation
NEB	Nudged elastic band
PES	Potential energy surface
PPTIS	Partial path transition interface sampling
PT	Parallel tempering
RC	Reaction coordinate

RE   Replica exchange
RRKM   Rice–Ramsperger–Kassel–Marcus theory
SDE   Stochastic difference equation
SM   String method
TI   Thermodynamic integration
TIS   Transition interface sampling
TPE   Transition path ensemble
TPS   Transition path sampling
TPT   Transition path theory
TS   Transition state
TSE   Transition state ensemble
TST   Transition state theory
US   Umbrella sampling

1   Rare Events in Complex Systems

During the past few decades computer simulation methods such as molecular dynamics (MD) and Monte Carlo (MC) simulation have grown into powerful and extremely versatile tools in theoretical condensed matter science. Today, these methods are run on fast computers to study complex systems consisting of up to millions of particles of interest in physics, materials science, chemistry, and biology with atomistic resolution. But despite the tremendous algorithmic advances and the steep increase in raw computing power that we have witnessed recently, many interesting and important processes still lie beyond the reach of current technology. The main reason for this limitation is that frequently the behavior of the system of interest is determined by phenomena occurring on vastly different time and length scales. The structure and dynamics of polymer solutions, for example, involves characteristic lengths ranging from the length of a chemical bond ($\sim$1Å) and the persistence length of a chain ($\sim$1 nm), to the extension of a coil ($\sim$10 nm), the inter-coil distance ($\sim$0.1 μm), and, finally, the size of the macroscopic sample ($\sim$1 cm) [1]. The time scales associated with the dynamics at the different levels of this hierarchy of length scales span an even wider range from the femtosecond regime of bond vibrations to the practically boundless characteristic times for coil motion near the glass transition [2].

One strategy for overcoming the difficulties associated with the wide range of length and time scales consists of coarse graining the description by eliminating unimportant variables and retaining only those degrees of freedom that are essential for the phenomena one wants to study. For example, the complexity of models for polymer solutions is often reduced using bead-spring models, in which single particles replace whole groups of atoms [2], or by systematically integrating out part of the microscopic degrees of freedom such that the total potential energy can be written in terms of effective interactions [3,4]. Such simplified models can be simulated very efficiently and the accessible time scales increase by orders of magnitude.

Development of such models, however, requires considerable insight into the nature of the problem (e.g., which degrees of freedom can be eliminated and which ones are essential) and such knowledge is often unavailable. For instance, it is unclear how one would go about coarse graining a supercooled liquid that approaches the glass transition. In this case, the collective variables capturing the essential physics of the process are unknown. This lack of a priori knowledge about mechanisms is very common, precluding the possibility of any systematic coarse graining procedure. Then, there is no way around a fully atomistic simulation, which, naturally, is complicated by the presence of widely disparate time scales. While no general solution to this problem exists, some progress has been made in the development of methods for the simulation of processes dominated by rare events. These methods and their applications are the subject of the present review article. In the following, we will be only concerned with systems in which atoms are described as classical particles. (Note, however, that the determination of the potential energy surface (PES) on which these atoms evolve may require quantum mechanical electronic structure calculations.)

Rare events are important if the system's dynamics consists of extended sojourns in long-lived stable states punctuated by rapid transitions between such states. (Here, we call a region of configuration space stable, if the system resides in it for a long time. Of course, there is a certain amount of arbitrariness in such a definition, but as long as there is a clear separation of time scales such a definition makes sense. In our terminology, "stable" also designates states that are thermodynamically metastable.) Examples of such processes include nucleation in first-order phase transitions, chemical reactions, transport phenomena in solids and liquids, biomolecular isomerizations or even transitions of comets between different orbits in the solar system [5]. Such transition events are rare, because the stable basins are separated from each other by high potential energy barriers or entropic bottlenecks. But while being rare, these transitions proceed swiftly when they occur. Consider, for instance, the autoionization of a water molecule in the bulk liquid. Here, the stable states between which the transition occurs are the intact water molecule H_2O and the dissociated fragments OH^- and H^+. From experiments we know that the average life time of a water molecule before the dissociation reaction occurs is of the order of ten hours for water under ambient conditions [6]. But when the reaction occurs, driven by a rare fluctuation of the solvent, the sequence of molecular events that lead to the formation of the separated ion pair takes place on a sub-picosecond time scale [7]. Thus, in this case there is a wide gap of characteristic times spanning more than 16 orders of magnitude. Time scale gaps of similar magnitude are also observed in the crystallization of a supercooled liquid close to coexistence, which proceeds via nucleation and growth. Since this process requires the formation of a free energetically unfavorable crystallite of sufficient size, the nucleation takes place rarely. But when it takes place, formation of the critical nucleus and subsequent growth happens quickly. (It is important to realize that in general the ascent to the top of the free energy barrier occurs as rapidly as the descent from it. This is simply required by the microscopic reversibility of the dynamics. Such barrier crossing events are just rare rather than slow!)

In principle, conventional computer simulations can be used to study the dynamics of processes involving rare events. One could, for instance, just follow the time evolution of the system with a MD simulation (using an empirical force field or first principle methods for the force calculation) and observe what happens. In such a case the time step required for a faithful simulation of the system is dictated by the shortest characteristic time present in the system, usually the femtosecond time scale of molecular bond vibrations and atomic collisions. Then, a very large number of such time steps would be required to observe even one single transition event. In simulating the autoionization of water with such a straightforward MD approach of the order of 10^{20} time steps of femtosecond length would be required to observe one single dissociation. The computational requirements of such a procedure are clearly beyond current capabilities (and will stay there for a while), but even in less extreme cases MD simulations of rare events are impractical (we would probably not call these events "rare" if it were otherwise).

Similar computational complications can occur in the calculation of structural (rather than dynamical) properties of complex systems. In MC simulations, for instance, an accurate determination of the equilibrium properties of a molecular system requires a proper sampling of all statistically relevant configurations. If, as in the above examples, configuration space is partitioned into stable basins by high (free) energy barriers, straightforward sampling with only local moves is unable to accomplish exhaustive sampling in the available CPU time. The reason is that to connect stable basins the system needs to traverse the low probability regions on the free energy barriers. In this case, the number of simulation steps required to move between adjacent stable regions by far exceeds the number of steps needed to obtain converged averages within each such region. If one is interested in structural properties only, which are mostly determined by the stable states rather than the transition regions, it is often possible to exploit the freedom of MC simulations and design moves that transport the system from one high likelihood region to another without passing through energetic and entropic bottlenecks. MC methods based on such smart moves have been developed and have led to huge increases in computational efficiency [8, 9]. Other recent approaches to find and/or identify stable states include simulated annealing [10], genetic algorithms [11, 12], hidden Markov models [13, 14], and basin hopping [15]. These methods, however, are outside the scope of this article. Here, we will focus on simulation methods designed to study the mechanistic and kinetic details of the transition processes themselves.

Often, at low temperature, stable states are associated with regions around single potential energy minima and transitions between these minima occur via saddle points of the PES. Saddle points are then transition states, i.e., mountain passes from which both stable states are equally accessible. In this case, a good strategy to explore stable states and transition routes between them is to search for stationary points on the PES, i.e., points such as minima and saddle points at which the gradient of the potential energy vanishes. Methods to do that are discussed in Sect. 2. The same section also includes a brief overview of the nudged elastic band method (NEB) [16], action-based methods [17], the coarse molecular dynamics (CMD) method [18], and the metadynamics method [19]. CMD and metadynamics are *one-ended* methods designed to explore the free energy landscape spanned by a few appropriately selected collective variables and to identify possible transition pathways to adjacent stable states. In contrast, the NEB method and the action-based methods are *two-ended* methods to determine pathways between two known stable states. Once transition states on the potential or on the free energy surface (FES) have been determined with these methods, the reaction kinetics can be studied with transition state theory (TST), a topic which is covered in Sect. 3. The central idea of this approach is to determine the flux through a so-called dividing surface separating the stable states between which the transition occurs. We will discuss various approaches based on this idea differing in how this flux is determined and which approximations are used to do that.

For systems in which the dynamics consist of long stays in the basins around potential energy minima interrupted by swift hops between them passing through saddle points, the assumptions of TST are often obeyed and the long time dynamics can be studied with the accelerated MD methods [20] discussed in Sect. 4. In these approaches, which include parallel replica dynamics, hyperdynamics, and temperature-accelerated dynamics, the rate of escape from the minima is artificially enhanced in one way or another. The natural dynamics on long time scales is then reconstructed from such boosted simulations and often dramatic speed-ups can be achieved.

While the methods considered in Sect. 2 are either static (transition state searches, NEB) or replace the true dynamics of the system with an artificial time evolution (metadynamics), the transition path sampling (TPS) method [21–23] presented and discussed in Sect. 5 deals with fully dynamical trajectories. On the basis of a statistical and reaction-coordinate free description of transition pathways, this methodology can be used to study the mechanism and the kinetics of transitions between known stable states. The central idea of this method is to first define the set of all reactive trajectories, i.e., the transition path ensemble (TPE) consisting of all trajectories that connect the stable states of interest, and then to sample these trajectories with MC methods acting in the space of trajectories, an idea that goes back to work of Pratt [24]. In the underlying statistical mechanics of trajectories reaction rate constants can be expressed in terms of path averages and various methods exist to evaluate these averages efficiently. As the TPS method does not require prior knowledge of a reaction coordinate and does not rely on the identification of particular features of the PES, it can be applied to rare transitions in complex systems with rugged PESs and/or entropic barriers. Owing to our own personal preferences and expertise, among all the methods treated in this review article TPS will be discussed in greatest detail.

A discrete version of TPS, in which pathways are viewed as sequences of adjacent potential energy minima, was developed by Wales and collaborators and is briefly discussed in Sect. 6.

Computational methods such as TPS yield sets of likely transition trajectories but do not automatically identify a small set of pertinent variables, ideally a one-dimensional reaction coordinate, that can be used to understand the system's dynamics in terms of a simplified low-dimensional model. Visual inspection of the trajectories with a molecular viewing program may sometimes help to single out the crucial variables, but, in general, for complex systems these variables remain hidden to the eye. Recently, a number of approaches have been developed to address the important problem of finding appropriate reaction coordinates. In all of these methods, the so-called committor (or commitment probability) plays a central role. As explained in Sect. 7, the committor is the probability that trajectories started from a given configuration end up in a particular stable state. Several committor-based analysis methods, including calculation of the transition state ensemble (TSE) and committor distributions [25, 26], Bayesian path statistics [27], genetic neural networks (GNN) [28], and likelihood maximization [29], are discussed in Sect. 8. The concept of the committor is seamlessly integrated into the string method [30–32] whose conceptual basis is provided by transition path theory (TPT) [33]. The key ideas of the string method are outlined in Sect. 9. In the following we will first define some basic concepts and then give a more detailed discussion of some of the methodologies mentioned above for identifying transition mechanisms and determine reaction rates.

2 Exploring (Free) Energy Landscapes

Structure and dynamics of classical mechanical systems are essentially determined by the total potential energy $V(r)$ usually given as a function of the coordinates $r = \{r_1, r_2, \ldots, r_{3N}\}$ of all N atoms. In molecular simulations, the potential energy is either modeled as an empirical potential or calculated directly from a solution of the electronic Schrödinger equation in the Born–Oppenheimer approximation. For typical condensed matter systems $V(r)$ is a complicated function with a multitude of minima, maxima, saddle points, and singular points. Often, the potential energy is pictured as a landscape in which the elevation in the z-direction corresponds to the value of the potential energy at a particular configuration r represented by a point in the xy-plane. An example of how one may imagine such a PES is shown in Fig. 1.[1] Although this suggestive perspective may assist our imagination, it is important to keep in mind that the landscape picture is a drastic simplification as the high-dimensional configuration space is represented by one or, at most, two dimensions.

Under certain circumstances the properties of condensed systems can be understood in terms of characteristic points of the PES [37, 38]. Transport in or on solids, chemical bond cleavage and formation, and reorganization processes in solid

---

[1] Figure 1 depicts the PES for one particle in the fluid phase of the Gaussian core model [34, 35]. In this system, particles interact via a Gaussian pair potential, $\phi(r) = \epsilon \exp(-r^2/\sigma^2)$, which accurately describes the effective interaction of polymer coils in solution [36]. The parameters σ and ϵ set the length and energy scales, respectively. The PES was calculated by translating one particle in a plane while keeping all other particles fixed at the positions of a typical configuration of the liquid phase at density $\rho = 0.25\,\sigma^{-2}$ and temperature $T = 0.01\,\epsilon/k_B$ for $N = 1,000$ particles.

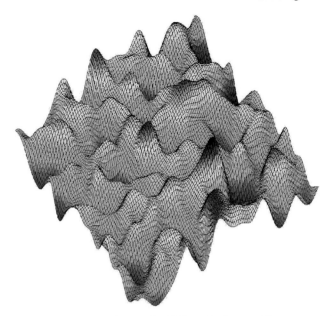

Fig. 1 Low-dimensional depiction of a complex PES as a landscape with numerous minima, maxima, and saddle points. The vertical elevation is given by the value of the potential energy and the xy-plane corresponds to the system's configuration space. At low temperatures, transitions between potential energy minima occur via saddle points that coincide with the mountain passes in this topographic perspective

clusters often belong to this class of processes. According to the canonical distribution

$$\rho(r) = \frac{1}{Z} \exp[-\beta V(r)], \tag{1}$$

valid for systems in contact with a heat bath at temperature T, configurations of low potential energy have higher probability to occur than those with high potential energy. In (1) $\rho(r)$ is the probability density for observing configuration r and $\beta = 1/k_B T$ is the reciprocal temperature. The configurational partition function

$$Z = \int dr \exp[-\beta V(r)] \tag{2}$$

normalizes the distribution of (1). (Note that, except in one dimension, also in microcanonical systems, i.e., systems evolving at constant energy, low-potential-energy configurations carry a larger statistical weight than high-potential-energy configurations.) At temperatures for which the thermal energy $k_B T$ is small compared to the potential energy barriers separating neighboring potential energy minima, the system will mainly fluctuate in the potential energy wells around the minima, the deep valleys of the PES. At these points, which correspond to the stable structures of the system, all first derivatives of the potential energy vanish and all second derivatives, i.e., the curvatures, are positive. Only rarely will a thermal fluctuation drive the system far enough from one minimum so that it can cross a potential energy barrier and move to an adjacent minimum. Because high-potential-energy configurations have small statistical weight, such a transition between minima is most likely to succeed if it crosses the potential energy barrier at its lowest point. This point is a first-order saddle point in which all first derivatives vanish and curvatures are positive in all directions but one. Wandering along this so-called unstable direction, the system can cross the barrier at the lowest possible energetic expense. In the presence of thermal noise the system does not pass exactly through the saddle point, or transition state (TS), but will cross the potential energy barrier somewhere close to it, provided the temperature is sufficiently low. In this low temperature case, both the equilibrium thermodynamics as well as the dynamics of the system can be deduced from knowledge of minima and saddle points and it is therefore desirable to have algorithms for finding these stationary points, i.e., points with vanishing gradients. Some of these methods will be briefly discussed in Sects. 2.1 and 2.2.

The landscape picture can be extended to systems at higher temperature, where entropic effects may become important, by introducing the concept of a *free energy landscape*. The definition of a meaningful free energy requires that a number of possibly collective coordinates $q(r) = \{q_1(r), q_2(r), q_3(r), \ldots, q_M(r)\}$, sufficient for a description of the essential physics of the problem, have been identified. The number M of selected collective variables is usually much smaller than the dimensionality of configuration space. Such collective variables, each of which is a unique function of the configuration r of the system, can, for instance, be dihedral angles of a biomolecule undergoing an isomerization reaction or the size of a crystalline cluster forming in an undercooled melt. The free energy as a function of the collective variables is then defined as the logarithm of a reduced partition function,

$$F(q_1, q_2, \ldots, q_M) = -k_B T \ln \int dr \exp[-\beta V(r)] \prod_{i=1}^{M} \delta[q_i - q_i(r)], \qquad (3)$$

where the integration extends over the whole configuration space. The free energy is related to the density distribution $P(q_1, q_2, \ldots, q_M)$ of the collective variables by

$$P(q_1, q_2, \ldots, q_M) \propto \exp[-\beta F(q_1, q_2, \ldots, q_M)]. \qquad (4)$$

The free energy $F(q)$ is the effective interaction between the collective variables that remains when all other degrees of freedom are integrated out. For the description of the statistics of the collective variables the exponential $\exp[-\beta F(q)]$ plays the same role the Boltzmann factor $\exp[-\beta V(r)]$ plays for the statistics of configurations. One can therefore apply the landscape perspective also to free energies as defined in (3) and view the surface depicted in Fig. 1 as an example of a FES. Minima in the free energy landscape then correspond to stable states in which the system is mainly observed. These stable states are separated by free energy barriers, which are crossed when the system performs a transition from one free energy minimum to another. It is, however, important to keep in mind that the free energy landscape is arbitrary in the sense that it strongly depends on the definition of the collective coordinates. Therefore, it does not make sense to speak about *the* free energy landscape as if the free energy was a unique inherent property of the system. Rather, free energy considerations are only meaningful if, at the same time, the collective variables $q(r)$ are specified. Nevertheless, provided these collective variables are appropriately defined, exploration of free energy landscapes can be very fruitful.

One way of exploring the free energy landscape is the parallel tempering method (PT) [39] (aka replica exchange (RE) [40, 41]). In PT/RE simulations one simultaneously runs many replicas of the system at different temperatures. At high temperature, the phase space is sampled efficiently, but at low temperature barriers are rarely crossed. A MC algorithm that occasionally exchanges replicas allows the crossing of high barriers while rigorously obeying the Boltzmann distribution for all temperatures. Subsequent application of (3) and (4) then yields the free energy for an arbitrary set of collective variables q. The PT/RE technique is thus an order-parameter-free method to determine important features of the free energy landscape of complex systems. A major drawback is that it does not give any information on the transition barriers, because these are not populated.

Established algorithms that *do* compute free energy barriers include umbrella sampling (US) [9, 42], thermodynamic integration (TI) [43], blue moon sampling [44], and the adaptive force method [45]. While these methods are extremely useful, they often rely on the definition of a single collective variable. In this review, we focus more on free energy methods that allow for multidimensional order parameters. An example of a computational technique to explore the free energy landscape efficiently using multiple collective variables is the metadynamics method [19] discussed in Sect. 2.5.

In the following sections, we will discuss several (but, due to space limitation, not all) techniques to explore (free) energy landscapes.

2.1 Saddle Point Search Algorithms

As the optimization of functions is a problem ubiquitous in many areas of science and technology, many numerical algorithms to find the location of local and global minima of functions have been developed and can be used for the exploration of the PES [46]. Methods for finding *local* minima can be roughly classified according to the highest-order derivatives they require. The simplex method, for instance, only necessitates evaluations of the energy itself. In other methods, such as the steepest descent and the conjugate gradient methods, first derivatives of the energy must be available. Finally, minimization procedures such as the Newton–Raphson algorithm or the Broyden–Fletcher–Goldfarb–Shanno algorithm make use of the second derivatives of the energy. As second derivatives are often not available (or only at a very high computational price) for molecular systems (for instance, in ab initio calculations), steepest descent and conjugate gradient are often the methods of choice.

Finding the lowest of all energy minima, i.e., the global energy minimum, is a much more challenging problem and several methods are available to address this important issue. As all these minimization algorithms have been described in detail in other places (see, for instance, [37] and [46]) and, furthermore, they are not central to this review article, we will not go into more detail on this topic here.

In general, finding saddle points on the PES is an even more complex problem than finding minima. The reason for this distinction is that a saddle point is a minimum in all directions except one, in which it is a maximum. So, saddle points cannot be found by simply walking uphill starting from a minimum. Rather, one has to walk uphill while at the same time staying close to the valley bottom. The eigenvector-following method is a formalization of this idea and allows one to efficiently find saddle points in complex molecular systems [47–49]. A detailed explanation of the eigenvector-following method including numerous applications to clusters, biomolecules, and glasses can be found in the book by Wales [37]. Methods such as eigenvector following, which start from a single minimum, are called single-ended methods. Other single-ended approaches are the activation-relaxation technique (ART) of Mousseau and Barkema [50–52], the dimer method [53] as well as the topological method of Tanase-Nicola and Kurchan [54, 55], in which a population of stochastically evolving walkers is steered towards dynamical bottlenecks by a birth/death process depending on the local stability properties of the PES. Double-ended methods, on the other hand, start the saddle point search from two points on different sides of the saddle point. Chain-of-states algorithms such as the NEB method discussed in the next section, are examples for such double-ended methods. Once the saddle point has been located using one of these methods, it is straightforward to determine the so-called minimum energy path (MEP) by walking downhill starting from the slightly perturbed saddle point. Note, however, that this minimum energy pathway is likely to differ from the dynamical path followed during the transition and does not even have to be similar to it.

2.2 Nudged Elastic Band Method

Another approach to determine transition states and minimum energy pathways consists of constructing a chain of states $\{r^{(0)}, r^{(1)}, \ldots, r^{(M)}\}$ that connects two known adjacent potential energy minima. Here, each state $r^{(i)}$ is a complete replica of the system's configuration. As illustrated schematically in Fig. 2, the chain of states acts like an elastic band with endpoints anchored in the minima. The central idea of most chain of states methods is to let the intervening states relax in a way that minimizes a particular object function $S[r^{(0)}, r^{(1)}, \ldots, r^{(M)}]$. This object function usually depends on the springs connecting adjacent states and on the underlying PES $V(r)$ and is not directly related to the dynamics of the system.[2] For an object function constructed

---

[2] If subsequent states in the chain of states are related by the rules of the natural underlying dynamics one should speak about "trajectories" rather than "chain of states." Methods based on trajectories, such as TPS, the string method, and action-based methods are discussed later in this review.

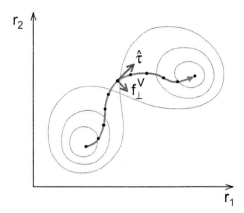

Fig. 2 Chain of states connecting two adjoining potential energy minima. In the NEB method, forces stemming from the springs connecting adjacent states act only in a direction tangential to the chain as described by the tangent vector $\hat{\tau}$. The forces $f_{\perp}^{V}$ caused by the PES act only in the directions orthogonal to $\hat{\tau}$

in a suitable way, the optimized chain will cross the potential energy barrier at the saddle point and join the potential energy minima following the MEP. The various existing chain-of-states methods differ in the particular form of the object function and in the way it is minimized [56–62]. Other methods to find saddle points also require knowledge of both potential energy minima they connect, but operate only with a pair of images rather than with a whole chain of states [63,64]. Since among the chain-of-states techniques the NEB method of Jonsson and collaborators [62] is particularly efficient and has been used in numerous applications, we will describe it in some detail here as an example. For a brief discussion of the relation of the NEB method to other chain-of-states methods we refer the reader to [16,62].

In the simplest form of the object function, each state $r^{(i)}$ is penalized with the potential energy $V[r^{(i)}]$ at that point and adjacent states are coupled with a harmonic spring,

$$S[r^{(0)}, r^{(1)}, \ldots, r^{(M)}] = \sum_{i=0}^{M} V[r^{(i)}] + \frac{k}{2} \sum_{i=1}^{M} [r^{(i)} - r^{(i-1)}]^2. \tag{5}$$

When this function is minimized while keeping the chain endpoints fixed, two effects compete with each other: on the one hand, the states along the chain tend towards the potential energy minima so that the sum of the potential energy terms becomes small; on the other hand, the harmonic springs try to keep the states approximately equidistant on a straight line to minimize the energy stored in the springs. For an appropriately selected force constant k, the optimized elastic band will cross the potential energy barrier near its lowest point (the transition state) and approximately follow the lowest energy band.

Such a straightforward approach is, however, plagued by two problems. First, the harmonic springs, which favor a straight path with regularly spaced images, tend to pull the path away from the MEP and the saddle point. This tendency to cut corners is particularly pronounced for large spring constants. Second, the images along the chain tend to slide down along the chain towards the potential energy minima under the action of the forces resulting from the underlying PES. This effect, which leads to a poor resolution of the chain in the important transition state region, is strongest for small spring constants. Often it is impossible to select a force constant k of the harmonic springs in a way that reduces both problems simultaneously to a satisfactory degree.

In the NEB method the "corner cutting" and "sliding down" problems are cured by *nudging* the elastic band. This is done by projecting out the normal component of the spring force and the parallel component of the physical force arising from the underlying potential. Thus, the total force f^{NEB} acting on image $r^{(i)}$ is given by

$$f^{\mathrm{NEB}}[r^{(i)}] = f_{\perp}^{V}[r^{(i)}] + f_{\parallel}^{s}[r^{(i)}], \tag{6}$$

where the normal physical force $f_{\perp}^{V}$ is given by

$$f_{\perp}^{V}[r^{(i)}] = -\nabla^{(i)}V[r^{(i)}] + (\nabla^{(i)}V[r^{(i)}] \cdot \hat{\tau})\hat{\tau} \tag{7}$$

and the parallel spring force reads

$$f_{\parallel}^{s}[r^{(i)}] = k\left[\left(r^{(i+1)} - 2r^{(i)} + r^{(i-1)}\right) \cdot \hat{\tau}\right]\hat{\tau}. \tag{8}$$

Here, the gradient $\nabla^{(i)}$ acts on the coordinates of image i only. The simplest way to estimate the tangent vector $\hat{\tau}$ required for the projection is

$$\hat{\tau} = \frac{r^{(i+1)} - r^{(i)}}{|r^{(i+1)} - r^{(i)}|}, \tag{9}$$

but better expressions for $\hat{\tau}$ exist [65]. Using the NEB-forces from (6), which completely eliminate the corner cutting and sliding down problems, one can relax the elastic band efficiently towards the MEP, for instance with simple steepest descent [62] or more sophisticated algorithms [62, 66]. With the climbing image nudged elastic band method (CI-NEB) [67], it is possible to locate the saddle point on the PES with high precision.

A NEB calculation usually requires only a few dozens of images and converges in a few hundreds of steps. Another computational advantage of the NEB method is that it is easy to implement on parallel computers and that it requires only first derivatives of the potential energy. The MEP and the transition state obtained by a NEB simulation are, however, purely static objects and the determination of kinetic properties requires combination of the NEB method with other approaches such as TST. The NEB method has been mainly applied, often in conjunction with harmonic

TST (see Sect. 3), to study numerous transport processes and reactions in and on solids [68–72] but also biomolecular isomerization processes [66, 73].

2.3 Action-Based Methods

The equations of motion of classical mechanics can be derived from variational principles such as Hamilton's principle of least action [74, 75]. This principle states that a physical path connecting a given initial configuration with a given final configuration in time τ makes the action

$$S = \int_0^\tau dt\, (K - V) \tag{10}$$

stationary. Here, K and V are the kinetic and the potential energy, respectively. Alternatively, Hamilton's equations of motion can also be obtained from the stationarity of

$$S = \int_0^\sigma ds\, \sqrt{2(E - V)}, \tag{11}$$

where the integral extends over the arc length s of the path of total length σ and E is the total energy of the system. In this so-called *Jacobi formulation* of the least action principle pathways are parametrized by arc length rather than time [75]. Also, this variational principle can be turned into an initial value problem with "equations of motion" in which time derivatives are replaced by derivatives with respect to arc length. While such variational principles embody a complete description of the system's dynamics, they are rarely used in practice and computational studies of the dynamics of molecular systems are mainly carried out using MD simulations based on the numerical integration of the equations of motion. For finding rare transition pathways connecting known configurations, however, a direct application of least action principles can be advantageous [17, 76].

The most straightforward way to implement the least action principle numerically consists of taking a global viewpoint, in which the whole trajectory, appropriately discretized, is considered at once. Accordingly, the integral of (10) [or of (11)] is replaced by a sum obtained from a suitable quadrature formula as suggested by Gillilan and Wilson [61]. Then, one searches for stationary points of the action in the high-dimensional space spanned by all coordinates along the discretized pathway. Such a direct numerical treatment is complicated by the fact that the stationarity condition of Hamilton's least action principle does not imply that physical pathways correspond to minima of the action as suggested by the (unfortunate) term "least action." Rather, physical pathways are, in general, located at saddle points of the action surface.[3] As mentioned above, finding saddle points is a burdensome task,

---

[3] As shown by Jacobi, physical pathways can never correspond to maxima of the action [77].

particularly in high-dimensional trajectory space. Therefore, this direct approach is usually unfeasible [78].

One approach suggested by Passerone and Parrinello [78,79] to circumvent these difficulties consists of imposing energy conservation along the discretized pathway by adding an appropriate term to the action. With suitable parameters, this additional term transforms the stationarity condition of the original least action principle into a true minimum condition. Since finding minima with methods such as steepest descent or conjugate gradient minimization is relatively undemanding [46], this modified method can be used to efficiently determine transition pathways in complex many-particle systems. To enhance the quality of the calculation and reduce its numerical cost, a Fourier representation of pathways is beneficial in this method [78, 80]. This way, paths are moved globally rather then locally resulting in more rapid convergence. This method has been applied to several rare-event-processes including the reorganization of a Lennard-Jones heptamer [78] and the isomerization of alanine dipeptide in the gas phase [79].

An alternative action-based method, named stochastic difference equation (SDE), has been devised by Elber and collaborators [17,76,81–83]. The central idea of this algorithm is to consider the errors of finite time step MD as statistically distributed and to use the distribution of these errors to write down a probability density function for approximate MD trajectories. In a sense, this method turns the imperfection of finite time step trajectories into an advantage by using the errors to formulate a convenient action functional. In the limit of small time steps this functional is the so-called Gauss action, which measures the mean square deviation of the dynamics from that prescribed by Newton's equations of motion. The Gauss action is identical to the so-called Onsager–Machlup action, introduced to study stochastic trajectories, in the case of vanishing friction [84]. Minimization of this action, either for pathways parametrized by time or arc length [83], yields a stable and efficient algorithm for determining pathways between known endpoints [85].

The fixed initial and final points of the trajectory provide additional stability such that in the SDE method a substantially larger time step than in straightforward MD simulations can be used [81, 86]. Indeed, it is the large time step that makes the SDE method powerful. As demonstrated in various applications including conformational transitions of glycosyltransferase [87], ion permeation through the gramicidin channel [88], the folding of protein A [89], the folding of cytochrome c [90], and the folding of a helical peptide [91], the accessible time scales can be extended to microseconds or even milliseconds, far beyond the time scales of typical MD simulations. In effect, by imposing fixed boundary conditions, the method excludes all trajectories that are numerically unstable. This essentially corresponds to filtering out high frequency motion such as bond vibrations, which are unimportant for large-scale molecular rearrangements. In activated processes, however, there are often high frequency motions related to rare but fast barrier crossing events. Since such high-frequency motions are also filtered out by the algorithm, action-based methods with large time steps are appropriate only for transitions involving slow diffusive barrier crossings rather than fast ballistic ones. In the latter case, methods such as TPS or the finite temperature string method described in Sects. 5 and 9, respectively, are more suitable.

Algorithmically, action-based methods are similar to the NEB method since in both cases a path functional is minimized. They differ, however, in the nature of the particular functional. While in the NEB method a path functional is constructed in an ad hoc way such that the path traverses the transition state separating reactants from products, the functional minimized in action-based methods corresponds, in principle, to the fully dynamical trajectories of classical mechanics. This property, however, is lost if extremely large time steps are used. In this case, the method yields a possible sequence of events that may be encountered by the system as it evolves from its initial to its final state, but a dynamical interpretation of such a sequence of states is not strictly permissible any more. Nevertheless, large time step trajectories that minimize the Gauss (Onsager–Machlup) action can provide possible scenarios for transitions that are computationally untreatable otherwise.

2.4 Coarse Molecular Dynamics

As explained above, the longest time scales accessible by MD simulations are limited by the short time steps dictated by the femtosecond time scale of the fastest atomic motions. In the CMD method of Hummer and Kevrekidis this limitation is overcome by considering the time evolution of the system in terms of a few "coarse" variables [18, 92]. These M variables $q = \{q_1(r), q_2(r), \ldots, q_M(r)\}$, each of which is a function of the microscopic degrees of freedom r, are supposed to capture the essential physics of the process of interest and therefore need to be selected with care. For molecular systems the coarse variables can, for instance, include dihedral angles, coordination numbers, solvent degrees of freedom, but also the geometry of the unit cell of a crystal. The basic idea of CMD, then, is to consider the dynamics only in the subspace spanned by the coarse variables and to effectively integrate out all fast degrees of freedom on the fly. The time derivatives needed for propagation of the coarse variables are determined from multiple short MD-runs carried out in the full phase space of the system. Accordingly, a CMD simulation consists of the following steps: for a given value of the coarse variables q many microscopic configurations r consistent with the particular values of q are constructed. This so-called lifting step is not unique and it can be carried out using biased MD simulations with an appropriate equilibration period. Starting from these initial conditions, short (unconstrained) MD trajectories in the full phase space are generated. The parameters characterizing the systematic and stochastic time evolution of the coarse variables are then determined by averaging over the trajectories. In this way one can estimate the underlying FES as a function of the coarse variables and the corresponding effective diffusion coefficient. These parameters are subsequently used to propagate the coarse variables in time. Assuming diffusive dynamics, analysis of this motion yields the free energy $F(q)$ and its minima and saddle points as well as the long-time behavior of the system.

Hummer and Kevrekidis have used the CMD method to study the free energetics and kinetics of the peptide fragment alanine dipeptide dissolved in water [18]. In this case, the dihedral angles ϕ and ψ were selected as the only coarse variables, which were propagated using multiple short MD runs of length 0.5 ps. From the time evolution of the coarse variables, the authors mapped out the FES and determined the rate constants for interconversions between different stable states. Other applications of the CMD method include the simulation of structural transitions in crystals [93] and of the filling and emptying of carbon nanotubes in water [94].

2.5 Metadynamics

A very similar perspective is adopted in the metadynamics method of Laio, Parrinello and collaborators [19, 95]. A recent account of metadynamics can be found in [96]. Also in this method, designed for exploring FESs and for finding pathways connecting stable states, the dynamics of a set of properly selected collective variables $q = \{q_1(r), q_2(r), \ldots, q_M(r)\}$ (the coarse variables of the CMD method) is considered. As in the CMD method, each of these variables is a function of the configuration r of the system.[4] In their lower-dimensional subspace, these variables evolve in time under the influence of the free energy $F(q)$. To prevent the system from getting trapped in free energy minima, a time-dependent bias potential $F_G(q,t)$ is added to the equation of motion. The bias potential is designed to mark the regions in the space of the collective variables q which have already been visited and to drive the system away from them, enabling it to escape free energy minima. This idea of a time dependent bias that destabilizes free energy minima is familiar from other computational techniques such as the flooding method [97–99], the local elevation method [100], or Wang–Landau sampling [101].

In the original version of metadynamics [19], called discrete metadynamics, the exploration of the free energy landscape $F(q)$ is driven by the forces

$$f_i(q) = -\frac{\partial F(q)}{\partial q_i}. \tag{12}$$

For a particular q, these forces can be evaluated as time averages in a MD simulation carried out in the full phase space of the system (including all microscopic degrees of freedom) with the collective variables $q(r)$ constrained at q. In contrast to the CMD method, where dynamical information is extracted from bursts of short MD trajectories, only the static thermodynamic driving force is determined in the constrained MD simulation of metadynamics. Using the forces $f_i(q)$ the collective

---

[4] In certain situations it is convenient to include other degrees of freedom into the configuration r in addition to the atomic coordinates. To study transitions between different structures of a crystalline material with metadynamics, for instance, the parameters specifying the shape of the simulation box are treated as dynamical variables.

variables could then be propagated, for instance, in a steepest descent way,

$$q_i^{t+1} = q_i^t + \delta q \frac{f_i^t}{|f_i^t|}. \tag{13}$$

Here, t numbers the time steps, q_i^t are the collective variables at time t and $f_i^t = f_i(q^t)$ is the force at q^t. The control parameter δq specifies the step size. Such a time evolution would, however, head towards the closest free energy minimum and get trapped there. To prevent this from happening, a repulsive Gaussian potential is deposited at every point visited in the space of the collective variables. Each of these potentials acts at all later times such that the total force at time t is given by the thermodynamic force plus a sum of the forces stemming from the Gaussian potentials,

$$\tilde{f}_i^t = -\frac{\partial F(q^t)_i}{\partial q_i} - w \frac{\partial}{\partial q_i} \sum_{t'<t} \exp\left(-\frac{|q-q^{t'}|^2}{2\sigma_q^2}\right). \tag{14}$$

The parameters w and σ_q set the strength and the width of the bias potential, respectively. By iterating the sequence of force calculations from a constrained MD simulation (or from a MC simulation) and the propagation step of (13) one obtains the time evolution of the collective variables. In the metadynamics terminology one then speaks of a *walker* traveling through the low-dimensional space of the collective variables on the FES $F(q)$.

The Gaussian potentials left behind by the walker yield a dynamics very different from that without this time dependent bias $F_G(q,t)$. The Gaussian potentials gradually fill any free energy minimum, such that the walker is forced to escape free energy minima and explore regions not visited before. This procedure permits a quantitative estimation of the underlying FES. Once all of the accessible space has been sampled in this way and all minima are filled up to nearly the same level, the bias $F_G(q,t)$ built up during the simulation approximately compensates the free energy such that, up to a constant, $F(q)$ is roughly given by the inverted bias potential,

$$F(q) \approx -\lim_{t\to\infty} F_G(q,t) = -w \sum_{t'<t} \exp\left(-\frac{|q-q^{t'}|^2}{2\sigma_q^2}\right). \tag{15}$$

The width σ_q determines the scale of the smallest features of the FES $F(q)$ that can be resolved. An increased resolution can be achieved by reducing σ_q albeit at the cost of slower sampling. Note that the above relation for the free energy has been introduced only heuristically. In contrast, the free energy calculation with the continuous version of metadynamics has been put on a rigorous formal basis [102].

As the walker leaves free energy minima most easily through saddle points in the free energy landscape, the metadynamics method is also capable of discovering new pathways between stable states and it can help to identify the respective mechanisms. Of course, the method is successful only if the space of the collective variables includes the degrees of freedom, which properly characterize the mechanism. Several improvements of the discrete metadynamics method as well as procedures to estimate the error and select the parameters are available [103].

The same general idea of a Gaussian bias that builds up in time and forces the system to explore new regions of configuration space was later incorporated directly into a MD simulation carried out in the full system including all microscopic degrees of freedom [95]. In this version of metadynamics, also known as continuous metadynamics forces originating from the bias potential formulated in terms of the collective coordinates act directly on the atomic coordinates. For this purpose, the time dependent forces

$$f_i^G(r,t) = -\frac{\partial}{\partial r_i} \frac{w}{\tau_G} \int_0^t dt' \exp\left(-\frac{|q(r) - q(r(t'))|^2}{2\sigma_q^2}\right) \qquad (16)$$

are added to the regular forces $f_i = -\partial V(r)/\partial r_i$ in the equations of motion of the system. Here, the parameter w/τ_G controls the strength of the Gaussian bias potential. The bias forces of (16), the calculation of which requires the derivatives of the collective variables with respect to the coordinates, can be easily incorporated in any MD algorithm including Car–Parrinello MD [104]. As in the case of the discrete version of metadynamics, the free energy $F(q)$ can be reconstructed from the accumulated bias potential. Under the assumption that the dynamics of the collective variables can be modeled by a Langevin equation, this relation can be proved rigorously and an error analysis can be carried out that guides the choice of the simulation parameters [102, 105]. Because of the artificial nature of the biased dynamics, the kinetics of the transformation can not be inferred directly from a metadynamics run. Rather, the information from such a simulation about the topography of the FES can be used as input for other approaches such as TST, TPS or the string method (all discussed below), which are capable of computing reaction rate constants. A recent improvement of metadynamics that combines the method with RE techniques has been proposed by Laio and coworkers [106, 107].

To date, metadynamics (mainly in its continuous variant) has been used to determine the FES and find possible transition pathways in numerous condensed matter systems. Applications include structural phase transitions [108, 109], chemical reactions [110–112], and biomolecular processes [113, 114].

3 Transition State Theory

In the previous section we have discussed computational methods for exploring potential and free energy landscapes. While these methods can help to identify the transition mechanism, they do not directly yield the rate at which these transitions occur. Once the relevant dynamical bottlenecks that oppose transitions between the stable states are determined, transition rates can be determined in the framework of TST [115–118]. The basic idea of this procedure goes back to Marcelin, Eyring, Horiuti, Polanyi, and Wigner [119–123], who realized the importance of transition states and developed the theory of chemical reaction kinetics. Since TST provides the basis for other rare event simulation methods and since the central concepts of the kinetics of rare events can be introduced in this context, we will discuss TST in some detail in this section.

3.1 Reaction Rate Constants

Consider a system with a bistable free energy $F(q)$ as shown in Fig. 3. Such a free energy might be characteristic for a molecule in solution undergoing isomerizations between two stable conformations A and B, i.e., for a unimolecular chemical reaction

$$A \rightleftharpoons B, \qquad (17)$$

in which molecules of type A and B are converted into each other. In this case, the collective variable (or reaction coordinate in the terminology of chemical reaction theory) might be an angle or bond length, which markedly differs in the two stable states. Imagine, now, that this system is followed for a long time and the value of q is recorded in regular intervals. (In experiments q might not be directly accessible, so let us just assume that the time evolution of q is followed in a computer simulation.) If the free energy barrier ΔF is high compared to the thermal energy $k_B T$, the barrier top, which must be crossed during transitions from A to B corresponds to low probability configurations. Accordingly, transitions from A to B and vice versa occur only rarely.

This gap between the time scale for motion in A and B and that for transitions between them is reflected in the trace $q(t)$ shown in the top panel of Fig. 4, which evidences how long stays in states A and B are interrupted by swift transitions between them. From a purely macroscopic point of view it makes sense to take a more coarse grained perspective and consider only the statistics of the transitions between A and B neglecting the detailed dynamics of $q(t)$ in the stable states. This can be done by following the time evolution of the indicator function $h_B[q(t)]$ as shown in

Fig. 3 Free energy $F(q)$ as a function of the reaction coordinate q. A free energy barrier of height ΔF located at q^* separates the stable states A and B. The *thin wiggly line* represents a trajectory going from A to B

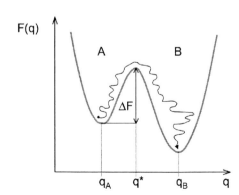

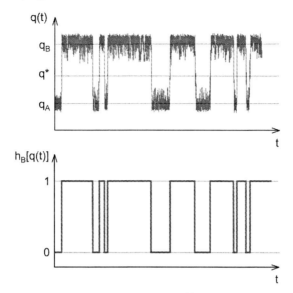

Fig. 4 *Top:* Time evolution of the reaction coordinate $q(t)$ for a system with a free energy $F(q)$ as shown in Fig. 3. Most of the time the reaction coordinate fluctuates around the values q_A and q_B typical for the stable states A and B, respectively. Rarely (on the time scale of the stable state fluctuations), the system switches between A and B. *Bottom:* Indicator function $h_B(t)$ for stable state B as a function of time t

the bottom panel of Fig. 4. This indicator function is defined such that it is unity, if the system is right of the barrier and 0 otherwise,

$$h_B[q(r)] = \begin{cases} 1 & \text{if } q(r) \geq q^*, \\ 0 & \text{if } q(r) < q^*. \end{cases} \tag{18}$$

The indicator function $h_A(q)$ for state A is defined analogously. The indicator functions h_A and h_B simply tell us whether the system resides in A or B.[5]

Using these indicator functions we can express the conditional probability to find the system in state B at time t provided it was in A at time 0,

$$C(t) \equiv \frac{\langle h_A[q(0)] h_B[q(t)] \rangle}{\langle h_A \rangle}. \tag{19}$$

Here, the angular brackets $\langle \cdots \rangle$ denote an average over equilibrium initial conditions, or, equivalently, a time average over a long trajectory, i.e.,

$$\langle h_A[q(0)] h_B[q(t)] \rangle = \lim_{\tau \to \infty} \frac{1}{\tau} \int_0^\tau dt' \, h_A[q(t')] h_B[q(t'+t)] \tag{20}$$

---

[5] For indicator functions defined according to (18), h_A and h_B are simply related by $h_B = 1 - h_A$ and definition of either h_A or h_B would suffice. Later, however, we will need regions A and B which are not complementary to each other and therefore we write all expressions using h_A *and* h_B such that they will remain valid later.

and

$$\langle h_{\mathrm{A}} \rangle = \lim_{\tau \to \infty} \frac{1}{\tau} \int_0^\tau \mathrm{d}t' \, h_{\mathrm{A}}[q(t')]. \tag{21}$$

The averages $\langle h_{\mathrm{A}} \rangle$ and $\langle h_{\mathrm{B}} \rangle$ are the probability to find the system in state A and B, respectively, in a long equilibrium trajectory.

The time correlation function $C(t)$ from (19) is a description of the transition statistics in the equilibrium system described in terms of the microscopic degrees of freedom. To make contact with a macroscopic description, appropriate for an experiment in which many molecules of type A and B are present in the sample, it is useful to consider the time evolution of the concentrations $c_{\mathrm{A}}(t) = N_{\mathrm{A}}(t)/V$ and $c_{\mathrm{B}}(t) = N_{\mathrm{B}}(t)/V$ defined as the number of molecules per volume V of type A and B, respectively. We imagine that the concentrations $c_{\mathrm{A}}(t)$ and $c_{\mathrm{B}}(t)$ can be determined experimentally in a time-resolved way. Since molecules can only convert into each other and are not created or destroyed, the total number of molecules $N = N_{\mathrm{A}}(t) + N_{\mathrm{B}}(t)$ as well as the sum $c_{\mathrm{A}}(t) + c_{\mathrm{B}}(t)$ are constant in time.

Consider now the case, where initially (at time 0) the system is prepared in a way such that all molecules are of type A. For a dilute solution in which molecules do not interact with each other, the concentration of molecules of type B as a function of time can be expressed in terms of the conditional probability $C(t)$ from (19) [124]:

$$c_{\mathrm{B}}(t) = \frac{N_{\mathrm{B}}(t)}{V} = \frac{N}{V} C(t) = (\langle c_{\mathrm{A}} \rangle + \langle c_{\mathrm{B}} \rangle) C(t), \tag{22}$$

where $\langle c_{\mathrm{A}} \rangle$ and $\langle c_{\mathrm{B}} \rangle$ are the equilibrium concentrations to which the system relaxes given enough time. Since the solution is dilute, the time evolution of the concentrations $c_{\mathrm{A}}(t)$ and $c_{\mathrm{B}}(t)$ should also be described well by the phenomenological rate equations

$$\begin{aligned} \dot{c}_{\mathrm{A}}(t) &= -k_{\mathrm{AB}} \, c_{\mathrm{A}}(t) + k_{\mathrm{BA}} \, c_{\mathrm{B}}(t), \\ \dot{c}_{\mathrm{B}}(t) &= k_{\mathrm{AB}} \, c_{\mathrm{A}}(t) - k_{\mathrm{BA}} \, c_{\mathrm{B}}(t), \end{aligned} \tag{23}$$

where k_{AB} and k_{BA} are the forward and backward reaction rate constants, respectively. These kinetic equations can be easily solved analytically and one finds that the deviations $\Delta c_{\mathrm{A}}(t) = c_{\mathrm{A}}(t) - \langle c_{\mathrm{A}} \rangle$ and $\Delta c_{\mathrm{B}}(t) = c_{\mathrm{B}}(t) - \langle c_{\mathrm{B}} \rangle$ from their respective equilibrium concentrations decay exponentially,

$$\Delta c_{\mathrm{A}}(t) = \Delta c_{\mathrm{A}}(0) \exp(-t/\tau_{\mathrm{rxn}}), \tag{24}$$
$$\Delta c_{\mathrm{B}}(t) = \Delta c_{\mathrm{B}}(0) \exp(-t/\tau_{\mathrm{rxn}}). \tag{25}$$

The reaction time τ_{rxn} is given in terms of the forward and backward reaction rate constants:

$$\tau_{\mathrm{rxn}}^{-1} = k_{\mathrm{AB}} + k_{\mathrm{BA}}. \tag{26}$$

It follows that for very different forward and backward reaction rate constants the reaction time is dominated by the larger one of the two. The solution of the rate equations expressed in (24) and (25) describes the way non-equilibrium concentrations relax towards equilibrium.

It follows from (25) that for the case considered above, where initially all molecules were of type A, the concentration of molecules of type B evolves as

$$c_B(t) = \langle c_B \rangle \left[1 - \exp(-t/\tau_{rxn}) \right]. \tag{27}$$

The phenomenological rate equations are expected to faithfully describe the kinetics of the system for times larger than the molecular time scale τ_{mol}, within which correlated barrier crossing events can take place, violating the assumptions of the kinetic equations. Roughly speaking, the molecular correlation time τ_{mol} is the time it takes for the system to forget how it got from A to B. For time t larger than τ_{mol}, any new event is independent of the previous barrier crossing. In this time regime, comparison of (22) and (27) yields

$$C(t) = \langle h_B \rangle (1 - \exp(-t/\tau_{rxn})). \tag{28}$$

Thus, also the time correlation function $C(t)$ is expected to approach its asymptotic value exponentially. This equality establishes a link between the microscopic dynamics of the system (left-hand side) and the phenomenological kinetic description in terms of reaction rate constants (right-hand side). If there is a separation of time scales, there will be a time regime $\tau_{mol} < t \ll \tau_{rxn}$ in which $C(t)$ grows linearly. It follows from (28) that in this regime

$$C(t) \approx k_{AB} t \tag{29}$$

or, equivalently, the time derivative of $C(t)$, also called the *reactive flux*, has a horizontal plateau with height equal to the forward reaction rate constant k_{AB} [125],

$$k(t) \equiv \dot{C}(t) \approx k_{AB}. \tag{30}$$

This equation is the basis for the estimation of reaction rate constants with TST. In the following sections we will discuss the various forms of TST that differ in how the rate constant is calculated.

3.2 TST Reaction Rate Constant

The condition $q(r) = q^*$ for the collective coordinate $q(r)$ defines a hypersurface in configuration space that separates states A and B, the so-called dividing surface. (For the time being we just assume that this dividing surface is appropriately chosen; in later sections we will discuss how to define this surface in an optimum way.) The goal of TST is to determine the flux through this surface. One can do that by calculating the time derivative of the correlation function $C(t)$. For stable states defined according to (18) we obtain

$$k(t) = \frac{\langle \dot{q}(0)\delta[q(0) - q^*]\theta[q(t) - q^*]\rangle}{\langle \theta[q^* - q(0)]\rangle}, \tag{31}$$

where $\theta(q)$ is the Heaviside step function. In deriving this equation we have exploited that $h_B(q) = \theta(q - q^*) = 1 - h_A(q)$. The above equation indicates that the flux $k(t)$ can be viewed as the average rate of change $\dot{q}$ of the reaction coordinate at the dividing surface defined by $q(r) = q^*$, with the additional condition that after time t the system is located in region B. This condition is violated for trajectories that, coming from A, make a brief excursion into B and return to A within τ_{mol} due to interactions of the reaction coordinate with other degrees of freedom. Such correlated recrossings [126] of the dividing surface lead to a short time transient behavior before the reactive flux approaches its plateau value k_{AB} [125].

The central approximation made in TST is to neglect recrossings of the dividing surface. This approximation corresponds to the assumption that all trajectories which cross the dividing surface heading into B will relax into B. Any subsequent crossing of the dividing surface is statistically independent. In this case, $\theta[q(t) - q^*]$ can be replaced by $\theta[\dot{q}(0)]$ and one obtains

$$k_{TST} = \frac{\langle \dot{q}\delta[q - q^*]\theta[\dot{q}]\rangle}{\langle \delta[q^* - q]\rangle} \times \frac{\langle \delta[q^* - q]\rangle}{\langle \theta[q^* - q]\rangle}, \tag{32}$$

where we have dropped the argument of q and $\dot{q}$ as the averages are over purely static quantities. The first factor on the right-hand side of the above equation is simply the average positive rate of change $\dot{q}$ of the reaction coordinate and the second factor is related to the free energy $F(q)$:

$$k_{TST} = \frac{1}{2}\langle |\dot{q}|\rangle_{q=q^*} \frac{e^{-\beta F(q^*)}}{\int_{-\infty}^{q^*} e^{-\beta F(q)}dq}. \tag{33}$$

Here, $\langle \cdots \rangle_{q=q^*}$ denotes an ensemble average with q constrained at q^*. This expression shows the exponential dependence of the reaction rate constant on the free energy difference between transition state and stable state. This dependence is characteristic for activated processes and is also known as Arrhenius behavior. From (33), which is the central result of TST, the following simple numerical procedure can be devised: first, calculate the free energy $F(q)$ as a function of the reaction coordinate $q(r)$ and determine the position q^* of the free energy maximum. Then, calculate the average positive flux $\langle |\dot{q}|\rangle_{q=q^*}$. In some cases, this average can be calculated analytically. If not, it can be determined, for instance, using a constrained MD simulation. Finally, combine these elements according to (33). The TST approximation of the reaction rate constant is computationally efficient, because it does not require the calculation of dynamical trajectories. However, TST can be shown [125] to always overestimate the reaction rate constant as indicated in Fig. 5.

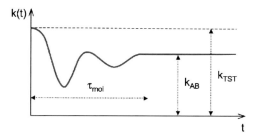

Fig. 5 Typical behavior of the reactive flux $k(t)$. For $t < \tau_{mol}$ correlated recrossings lead to a reduction of $k(t)$ with respect to $k(0)$ before it settles on a plateau whose height equals the forward reaction rate constant k_{AB}. The TST approximation k_{TST} of the reaction rate constant equals $k(0)$, the reactive flux at time $t = 0$

3.3 Harmonic Approximation

The calculation of reaction rate constants in TST can be further simplified if the stable states correspond to single potential energy minima at r_A and r_B and if during transitions the potential energy barrier is crossed near a saddle point r_{TS} in the PES as indicated in Fig. 6. In this version of TST these stationary points need to be exactly located. At low temperatures one can then approximate the PES near the stable states and the saddle points with a Taylor expansion truncated after the quadratic term. In this harmonic approximation, all averages appearing in (32) can be calculated analytically. To do that one first carries out a normal mode analysis in the potential energy minimum A as well as in the TS by diagonalizing the mass weighted Hessian

$$H_{ij} = \frac{1}{\sqrt{m_i m_j}} \frac{\partial^2 V(r)}{\partial r_i \partial r_j}, \tag{34}$$

where m_i is the mass associated with degree of freedom r_i. In the potential energy minimum A, this normal mode analysis yields eigenvalues λ_i^A which are all positive. (Eigenvalues that vanish due to the translational and rotational invariance of the Hamiltonian are disregarded.) The eigen-frequencies in the stable state are then given by the square root of these eigenvalues, $\omega_i^A = \sqrt{\lambda_i^A}$. In contrast, at the saddle point exactly one of the eigenvalues λ_i^{TS} is negative while all others are positive. The normal mode corresponding to the negative eigenvalue is the so-called unstable mode and it is in this direction that the system is assumed to cross the transition state in the harmonic approximation. (The direction of the unstable mode is denoted by a red arrow in Fig. 6.) Accordingly, the dividing surface is the plane normal to the direction of the unstable mode. The frequencies at the transition state in directions orthogonal to that of the unstable mode are given by $\omega_i^{TS} = \sqrt{\lambda_i^{TS}}$.

Using the normal modes that diagonalize the mass weighted Hessian both the expressions $\langle \dot{q}\delta[q - q^*]\theta[\dot{q}]\rangle$ and $\langle\theta[q^* - q]\rangle$ appearing in (32) can be easily calculated analytically as canonical averages, yielding the TST-rate constant k_{TST}^{ha} in the harmonic approximation,

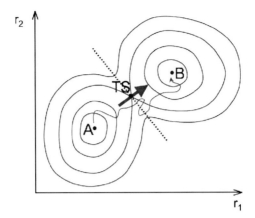

Fig. 6 Two-dimensional PES depicted with lines of equal potential energy. Here, the stable states A and B correspond to minima in the PES and the transition state TS that needs to be crossed during a transition between the stable states is a saddle point on the PES. The *thick arrow* indicates the direction of the unstable mode and the *dashed line* is the planar dividing surface orthogonal to this direction. The reactive trajectory shown in red recrosses the dividing surface twice

$$k_{\text{TST}}^{\text{ha}} = \frac{1}{2\pi} \frac{\prod_{i=1}^{n} \omega_i^{\text{A}}}{\prod_{i=1}^{n-1} \omega_i^{\text{TS}}} \exp(-\beta \Delta V), \tag{35}$$

where n is the number of the non-vanishing eigenvalues in the minimum and $\Delta V = V(r_{\text{TS}}) - V(r_{\text{A}})$ is the potential energy difference between saddle point and minimum. Thus, the reaction rate constant in the harmonic approximation can be written in the Arrhenius form as the product of two factors,

$$k_{\text{TST}}^{\text{ha}}(T) = \nu \exp(-\beta \Delta V), \tag{36}$$

where we have included the temperature T as an argument of k_{TST} to emphasize the temperature dependence of the reaction rate constant. The prefactor ν, depending on the frequencies in the minimum and at the saddle point, is an entropic factor which is large if the passageway at the saddle point is wide and small if it is narrow. While this prefactor can play a role in some cases, it is usually the exponential dependence on ΔV which is the dominating factor.

3.4 RRKM Theory

In the previous section we have determined the rate constant of a thermal system at temperature T. Often, however, it is more appropriate to consider systems not coupled to a heat bath and evolving at constant energy E. Also in this case the harmonic approximation is applicable, but the averages of (32) are to be taken over the microcanonical ensemble

$$\rho(x) = \delta[E - \mathcal{H}(x)]/g(E), \tag{37}$$

where

$$g(E) = \int \mathrm{d}x\, \delta[E - \mathcal{H}(x)] \tag{38}$$

is the density of states, $\mathcal{H}(x)$ is the total energy of the system and x denotes a state in phase space including positions and momenta. The analytical calculation of these microcanonical averages does not pose any problem and one obtains for the reaction rate constant:

$$k_{\mathrm{RRKM}}(E) = \frac{1}{2\pi} \frac{\prod_{i=1}^{n} \omega_i^{\mathrm{A}}}{\prod_{i=1}^{n-1} \omega_i^{\mathrm{TS}}} \left[\frac{E - V(r_{\mathrm{TS}})}{E - V(r_{\mathrm{A}})} \right]^{n-1}. \tag{39}$$

This is the Rice–Ramsperger–Kassel–Marcus (RRKM) expression of the reaction rate constant at energy E [127]. For total energies E below the potential energy of the transitions state, i.e., for $E < V(r_{\mathrm{TS}})$, $k(E)$ vanishes. The canonical and the microcanonical rate constants are related by a Laplace transform,[6]

$$k(T) = \frac{\int \mathrm{d}E\, k(E) g(E) \exp(-\beta E)}{\int \mathrm{d}E\, g(E) \exp(-\beta E)}, \tag{40}$$

and for systems with many degrees of freedom they are identical if one identifies the temperature of the microcanonical system as $\beta = \partial \ln g(E)/\partial E$. RRKM theory has been successfully used to calculate reaction rate constants for chemical reactions in the gas phase, for instance for proton transfer in small protonated water clusters [128].

3.5 Variational Transition State Theory

The application of TST and its more sophisticated improvements hinges on the definition of a suitable reaction coordinate capable of capturing the essence of the underlying mechanism. As explained above, in harmonic TST the reaction coordinate is identified with the direction of the unstable mode at the transition state. In this case, the dividing surface separating the stable states from each other is taken to be a hyperplane orthogonal to this direction. When non-linear effects are important such an approach may fail and the dividing surface must be found in a different way. In some cases a physically motivated choice of a dividing surface may yield satisfactory results.

---

[6] Note that this is true in general and not only for TST reaction rate constants in the harmonic approximation.

Alternatively, a variational principle provides a more systematic guide for finding a good dividing surface [121, 129–131]. This principle is based on the fact that the rate constant obtained from TST is always larger than the true rate constant. Hence, the optimum dividing surface is the one that minimizes the corresponding rate constant. Jónsson and collaborators have used this principle to derive simulation algorithms for finding the optimum hyperplanar dividing surface [132]. Further improvements may be achieved using more general dividing surfaces expressed in terms of curvilinear coordinates such as bending and torsional angles [133].

3.6 Dynamical Corrections

In TST, neglecting dynamical recrossings of the dividing surface leads to an overestimation of the transition rate constant. A procedure to calculate appropriate corrections was suggested by Yamamoto [134] and Keck [135] and was further developed by Bennett [136] and Chandler [125]. In this approach, dynamical trajectories initiated at the dividing surface provide the necessary information. The procedure is based on the exact calculation of the reactive flux $k(t)$ using a particular factorization of (31):

$$k(t) = \frac{\langle \dot{q}(0)\delta[q(0) - q^*]\theta[q(t) - q^*]\rangle}{\langle \delta[q^* - q(0)]\rangle} \times \frac{\langle \delta[q^* - q(0)]\rangle}{\langle \theta[q^* - q(0)]\rangle}. \tag{41}$$

This exact expression is similar to that of (32), but includes the full time dependence of $k(t)$. Written in a more suggestive way, the above equation reads

$$k(t) = \langle \dot{q}(0)\theta[q(t) - q^*]\rangle_{q=q^*} \frac{e^{-\beta F(q^*)}}{\int_{-\infty}^{q^*} e^{-\beta F(q)}dq}. \tag{42}$$

The second factor on the right-hand side of the above equation, which appears also in the TST rate constant, depends on the free energy difference between transition state and stable state and can be calculated using standard free energy estimation methods. The first factor is an average over dynamical trajectories started from configurations on the dividing surface, generated, for instance, using constrained MD simulations [44]. For any particular initial condition, the rate of change $\dot{q}(0)$ of the reaction coordinate at time 0 contributes to the average only if the system is located in state B a time t later. Multiplication of the two factors yields $k(t)$ and the reaction rate constant k_{AB} is then given by its plateau value as illustrated in Fig. 5.

It is practical to express the reaction rate constant k_{AB} as the product of the TST reaction rate constant and the *transmission coefficient* κ that corrects for the dynamical recrossings:

$$k_{AB} = \kappa k_{TST}. \tag{43}$$

Accordingly, the transmission coefficient equals the plateau value of the function

$$\kappa(t) = \frac{\langle \dot{q}(0)\theta[q(t) - q^*]\rangle_{q=q^*}}{\langle \dot{q}(0)\theta[\dot{q}(0)]\rangle_{q=q^*}}. \tag{44}$$

In the absence of recrossings of the dividing surface, the transmission coefficient is unity, $\kappa = 1$, while it is smaller than one if recrossings occur. While both the TST reaction rate constant k_{TST} and the transmission coefficient κ strongly depend on the location of the dividing surface, their product, i.e., the reaction rate constant k_{AB} does not. The efficiency of the reactive flux calculation, however, depends sensitively on the properties of the dividing surface and it is optimal for a dividing surface minimizing the number of recrossings. For a large number of recrossings, the straightforward calculation of the transmission coefficient according to (44) becomes inefficient. A number of improved algorithms for estimating κ have been put forward recently [117, 137, 138]. Some issues related to the optimum dividing surface and the estimation of the transmission coefficient are discussed in [118]. Recently, dynamical systems theory has been used to develop a phase space formulation of TST, in which the dividing surface is free of local recrossings [139–142]. How useful this approach will be for complex systems with many stationary points is an open question.

4 Accelerated Molecular Dynamics

The term *accelerated molecular dynamics* refers to a set of simulation methods designed to study the long-time basin hopping dynamics in systems dominated by energetic effects (as opposed to entropic effects), for instance transport processes in and on solids at low temperatures. In such systems long stays in the potential energy basins near minima are interrupted by rapid transitions from one basin to another. By artificially encouraging basin escape, these methods achieve high efficiency increases with respect to simple MD simulation. While the introduced bias corrupts the short time motion within the potential energy basins, proper corrections based on TST make sure that the relative probabilities for escape through different routes remain unchanged. Therefore, accelerated MD methods yield the correct sequence of state-to-state transitions, thus reproducing the long-time dynamics of the system. The main advantage of these methods, when applicable, is that they do not require any prior knowledge of possible transition routes. Rather, the system itself finds its way through the network of potential energy basins. A more coarse-grained view is taken in the related kinetic Monte Carlo (KMC) method, in which no short time dynamics is carried out at all [143, 144]. Instead, transitions between stable conformations are executed at random according to the respective reaction rates. While in early KMC work the list of possible transitions was established in advance, more recently methods for finding these transitions and computing their reaction rate constants on the fly have been put forward [145]. In the following, we will discuss various accelerated MD methods, all developed by Voter and collaborators. For a recent review on this topic we refer the reader to [20].

4.1 Parallel Replica Dynamics

In the parallel replica dynamics method the time scale of MD simulations is extended by distributing the computation on many processors in a way that requires only little information exchange between them yielding almost linear speed-up in many cases. This method is applicable whenever the distribution of escape times t from a potential (or free) energy basin is exponential,[7]

$$p(t) = k\exp(-kt), \tag{45}$$

where k is the rate constant for basin escape. The parallel replica method is particularly simple to implement and the most accurate accelerated method since it does not rely on the TST assumption of no recrossings of the dividing surface.

A parallel replica simulation is carried out as follows. First, M identical copies of a certain configuration, the replicas, are distributed on M processors. On each of them, an MD simulation is run with randomization of the momenta for a short dephasing time τ_{deph} in order to eliminate correlations between the replicas. After that, the MD simulations on all processors are run independently in parallel until the first trajectory escapes the potential energy basin. (Basin escape can, for instance, be detected by periodical steepest descent minimization. If such a minimization converges to a different minimum than the previous one, a transition from one basin to another must have happened between minimizations.) Once a transition has occurred on one of the processors, all simulations are stopped and the clock is advanced by the sum of the times elapsed on all processors. Finally, the trajectory which has escaped the minimum is propagated for another short time τ_{mol} to permit for possible correlated recrossings into the original basin (the total time must be adjusted accordingly). The final configuration then serves as the starting point for a new iteration of the whole procedure.

It can be shown [147] that for an exponential distribution of escape times the way of time keeping described above yields the correct state-to-state dynamics even if processors with varying speeds are used. Provided that the reaction time per processor is long compared to the sum of dephasing time and recrossing time, $1/kM \gg \tau_{deph} + \tau_{mol}$, a parallel replica simulation carried out on M nodes leads to an ideal M-fold increase in computing speed. This feature of the parallel replica dynamics method has been exploited to study folding pathways of small proteins on up to thousands of computers in parallel [148]. Other applications include the simulation of chemical reactions [149], the growth of clusters of silicon interstitials [150], and hydrogen diffusion in solids [151]. Recently, the parallel replica dynamics method has been generalized to non-equilibrium situations with a time-dependent rate constant for escape and has been applied to follow the time evolution of a strained carbon nanotube [152].

---

[7] The parallel replica formalism has been generalized to non-exponential escape time distributions. In this case, the calculation of the advanced time becomes more involved [146].

4.2 Hyperdynamics

The central idea of the hyperdynamics method is to carry out a molecular dynamics simulation of the system on a PES $\tilde{V}(r)$ modified by the addition of a non-negative bias potential $\Delta V(r)$ [153, 154]:

$$\tilde{V}(r) = V(r) + \Delta V(r). \tag{46}$$

The hyperdynamics method is applicable to systems that obey the assumptions of TST: there are no recrossings of the dividing surface or correlated hopping events. (These assumptions need to be satisfied also on the modified surface.) The bias $\Delta V(r)$ is constructed to vanish on the potential energy barriers and to have a finite value only in the basins around the minima. Such a bias lifts the bottom of the minima with respect to the transition states thus effectively lowering the height of the potential energy barriers and facilitating escape from the basin. Since the PES remains unchanged at the transition states, the relative escape rates through different exit routes are the same with and without bias. Therefore, a hyperdynamics simulation yields a realistic sequence of state-to-state hops while the short-time dynamics within the minima is sacrificed by the bias potential. From the bias potential sampled during the simulation, the overall boost, i.e., the factor by which the dynamics is accelerated with respect to a regular MD simulation, can be determined as an average over the trajectory on the modified PES [153]:

$$\frac{t_{\text{hyper}}}{t_{\text{MD}}} = \langle \exp(\beta \Delta V) \rangle. \tag{47}$$

Thus, for a given bias, the boost decreases with increasing temperature.

The construction of a computationally inexpensive bias function $\Delta V(r)$ that yields large boost factors is a non-trivial matter. One approach to construct bias potentials is based on the observation that near a potential energy minimum all eigenvalues of the Hessian matrix are positive while there is exactly one negative eigenvalue at saddle points. Accordingly, one can define a bias potential that is positive where the smallest eigenvalue of the Hessian is positive and zero elsewhere [153]. Various improvements based on this general idea have been proposed [20]. Other simpler bias functions are possible, but they lead to smaller speed-ups. A comparison of different bias potentials can be found in [20]. Depending on the system, boost factors of up to 10^5 have been obtained in hyperdynamics simulations, for instance in the simulation of vacancy diffusion on surfaces [155]. Among other applications, the hyperdynamics method has been used to study biomolecular systems [156], the thermal desorption of n-alkanes [157], and the dynamics of adatoms [158].

4.3 Temperature-Accelerated Dynamics

Also in the temperature-accelerated dynamics method the rate of basin escapes is artificially increased [159]. In this case this is done by running a MD simulation at a temperature T_{high} which is chosen higher than the temperature T of interest. At the higher temperature, transitions occur more frequently, though the basin escapes might occur through different routes than they would at the temperature T. To reconstruct the right sequences of state-to-state hops from the high temperature trajectory, one proceeds as follows. Whenever a transition from one basin to the next is detected in the high temperature trajectory, the simulation is stopped and the nearest saddle point is located (for instance with the NEB method described in Sect. 2.2). The trajectory is reflected back into the basin it was about to leave and the simulation is carried on. All attempted escapes are treated in this way. This procedure is continued to a maximum time, for which a rigorous criterion exists [160]. From the sequence of attempted escape events and the respective saddle point energies one can then determine which transition would have happened first at the temperature of interest. This transition is then carried out.

Temperature-accelerated dynamics relies on the validity of harmonic TST for the basin escape both at T_{high} and T. As a consequence, the high temperature T_{high}, and hence the boost factor, are controlled by the height of the lowest energy barrier for escape. The speed-ups obtainable with the temperature-accelerated method often exceed those of the other accelerated dynamics methods. In simulations of vapor deposited crystal growth, for instance, boost factors of about 10^7 have been achieved extending the accessible time range to the second regime [161]. The temperature-accelerated dynamics method has been mainly used to study transport processes in and on solids [162, 163].

5   Transition Path Sampling

TPS is a set of computational techniques to study the mechanism and the kinetics of rare transitions occurring in complex systems. Its application is particularly appropriate for complex systems in which the initial state A, such as the reactants of a chemical reaction, as well as the final state B, the products, are known, but the reaction mechanism is unknown. In this section we will first describe in detail the TPS formalism and then discuss the salient capabilities as well as the limitations of TPS. For more information on the formalism and the applications of TPS we refer the reader to several review articles covering various aspects of TPS at different levels of detail [124, 164–169].

The basic situation that can be addressed with the TPS method is illustrated in Fig. 7. Here, the stable states A and B correspond to regions in configuration space (or, more generally, in phase space) characterized in terms of the microscopic variables. Each of these regions is stable in the sense that if the system is initialized in the region it will most likely stay inside it for a long time. The two regions

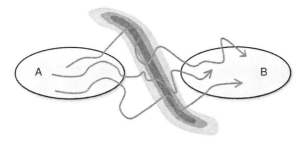

Fig. 7 The regions A and B in configuration (or phase) space correspond to stable states in which the system resides most of the time. Only dynamical trajectories that start in stable region A and end in stable region B have a weight in the TPE that differs from zero

A and B are separated by an unknown and possibly rough barrier. The dynamics of the system can be deterministic or stochastic, but in any case it is supposed to be Markovian, i.e., the probability of the future time evolution is fully determined by the current microscopic state of the system and does not depend on prior microscopic states. Many different kinds of dynamics belong to this class, including deterministic dynamics such as Newtonian dynamics and thermostated Nose–Hoover dynamics, as well as stochastic dynamics such as Langevin dynamics and MC "dynamics". If the system is initially placed in, say, A it will eventually cross the free energy barrier and move into the other stable state, B. In general, many different routes that carry the system from A to B are available and the search for the "typical" pathway is meaningless. Instead, the goal of a TPS simulation is to find all transition pathways and to extract mechanistic and kinetic information from them.

5.1 Transition Path Ensemble

The basis of the TPS method is the definition of the ensemble of all reactive trajectories, i.e., the ensemble of pathways that start in A at time 0 and end in B at time $\mathcal{T}$ later. (We will return later to the problem on how to specify the time $\mathcal{T}$.) Each trajectory, or path, is represented by an ordered sequence of microscopic states:

$$x(\mathcal{T}) \equiv \{x_0, x_{\Delta t}, x_{2\Delta t}, \ldots, x_{\mathcal{T}}\}. \tag{48}$$

Here x_t denotes the complete microscopic state of the system at time t (x_t is often called the time slice at time t). Depending on the dynamics, x consists of the positions and momenta of particles, $x = \{r, p\}$, or of the positions $x = \{r\}$ only. Consecutive states along the trajectory are separated by a time increment Δt. Such a trajectory could, for instance, result from carrying out $L = \mathcal{T}/\Delta t$ MD steps.

Since the underlying dynamics is assumed to be Markovian, the probability density for observing a particular L-step trajectory can be written as the product of the distribution $\rho(x_0)$ of the initial conditions x_0 with the product of all single time step transition probabilities:

$$P[x(\mathcal{T})] = \rho(x_0) \prod_{i=0}^{\mathcal{T}/\Delta t - 1} p(x_{i\Delta t} \to x_{(i+1)\Delta t}). \tag{49}$$

Naturally, the specific form of the single time step transition probabilities $p(x_t \to x_{t+\Delta t})$ depends on the particular type of dynamics one considers. Similarly, the distribution of initial conditions must be appropriate for the situation under consideration.

The path probability specified in (49) is the probability density for finding a particular trajectory without any condition on where the path begins and where it ends. In the TPS method one is, however, interested in a small set of pathways only, namely those that connect A with B. In order to restrict the path ensemble accordingly we assign a weight of 0 to all pathways that do not connect A with B. We can do that by multiplying the path probability of (49) with the characteristic functions of regions A and B acting on the initial and final point of the trajectories, respectively:

$$\mathcal{P}_{AB}[x(\mathcal{T})] \equiv h_A(x_0)\mathcal{P}[x(\mathcal{T})]h_B(x_\mathcal{T})/Z_{AB}(\mathcal{T}). \tag{50}$$

Here, the functions $h_A(x)$ and $h_B(x)$ are defined such that they are unity if the argument is located in the respective region and vanish otherwise,

$$h_A(x) = \begin{cases} 1 & \text{if } x \in A, \\ 0 & \text{if } x \notin A, \end{cases} \tag{51}$$

and $h_B(x)$ is defined analogously. While the specification of the initial and the final state is not always a trivial task (the native and denatured states of a protein are a point in case), one can usually study the properties of A and B with straightforward MD simulation and find a suitable description of them in terms of the microscopic degrees of freedom. We will discuss this point in more detail below. The "path partition function"

$$Z_{AB}(\mathcal{T}) \equiv \int \mathcal{D}x(\mathcal{T}) \, h_A(x_0)\mathcal{P}[x(\mathcal{T})]h_B(x_\mathcal{T}) \tag{52}$$

is a factor that normalizes the path distribution of (50). Here, the notation

$$\int \mathcal{D}x(\mathcal{T}) \equiv \int \cdots \int dx_0 dx_{\Delta t} dx_{2\Delta t} \cdots dx_\mathcal{T} \tag{53}$$

implies an integration over all time slices. Interestingly, the path partition function $Z_{AB}(\mathcal{T})$ equals the probability that an arbitrary trajectory of length $\mathcal{T}$ starts in A at time 0 and ends in B at a time $\mathcal{T}$ later. This fact can be used to derive algorithms for the calculation of reaction rate constants from the reversible work required to manipulate ensembles of trajectories [170, 171].

The path probability spelled out in (50) is a complete statistical description of all pathways connecting A with B within time $\mathcal{T}$. This set of pathways together with the weight of (50) is called the transition path ensemble (TPE). Pathways sampled from the TPE according to their statistical weight can be analyzed (subsequently or on the fly) to yield information about the details of the transition mechanism.

To construct the TPE for a specific process one has to use the appropriate distributions of initial conditions and short time transition probabilities. For an equilibrium system in contact with a heat bath, the distribution of initial conditions is canonical, while for an isolated equilibrium system at constant energy the initial conditions are distributed microcanonically.

The short time transition probabilities appearing in the trajectory weight are determined by the underlying dynamics. For a classical mechanical system evolving according to Hamilton's equations of motion,

$$\dot{r} = \frac{\partial \mathcal{H}(r, p)}{\partial p}, \qquad \dot{p} = -\frac{\partial \mathcal{H}(r, p)}{\partial r}, \tag{54}$$

the time evolution is deterministic such that the microscopic state x_0 of the system at time 0 is mapped onto a unique state x_t at time t,

$$x_t = \phi_t(x_0). \tag{55}$$

Accordingly, the short time transition probability is given by a Dirac delta function without any stochastic spread,

$$p(x_t \rightarrow x_{t+\Delta t}) = \delta\left[x_{t+\Delta t} - \phi_{\Delta t}(x_t)\right]. \tag{56}$$

In this case, the transition path probability is simply given by a product of delta functions,

$$\mathcal{P}_{AB}[x(\mathcal{T})] = \frac{\rho(x_0)}{Z_{AB}(\mathcal{T})} h_A(x_0) \prod_{i=0}^{\mathcal{T}/\Delta t - 1} \delta\left[x_{(i+1)\Delta t} - \phi_{\Delta t}(x_{i\Delta t})\right] h_B(x_{\mathcal{T}}), \tag{57}$$

where

$$Z_{AB}(\mathcal{T}) = \int dx_0 \, \rho(x_0) h_A(x_0) h_B(x_{\mathcal{T}}). \tag{58}$$

(Because of the properties of the Dirac delta function all integrations except the one over x_0 can be trivially carried out analytically.) The path ensemble from (57) is equally valid for other types of deterministic dynamics such as thermostated Nose–Hoover dynamics or Gaussian isokinetic dynamics.

If the system evolves stochastically, the short time transition probability is spread out rather than singular. For instance, in the case of Brownian dynamics the time evolution is given by [172, 173]

$$m\gamma\dot{r} = -\frac{\partial V(r)}{\partial r} + \mathcal{F},$$ (59)

where $\mathcal{F}$ is a delta-correlated Gaussian random force with zero mean and a variance given by the fluctuation–dissipation theorem:

$$\langle \mathcal{F}(t)\mathcal{F}(0)\rangle = 2m\gamma k_{B}T\delta(t).$$ (60)

Then, the corresponding short time transition probability is Gaussian,

$$p(r_t \rightarrow r_{t+\Delta t}) = \frac{1}{\sqrt{2\pi\sigma^2}} \exp\left\{ -\frac{\left(r_{t+\Delta t} - r_t + \frac{\Delta t}{\gamma m}\frac{\partial V}{\partial r}\right)^2}{2\sigma^2} \right\},$$ (61)

with variance

$$\sigma^2 = \frac{2k_{B}T}{m\gamma}\Delta t.$$ (62)

The finite width of the transition probability is a consequence of the random character of the motion. Appropriate transition probabilities for other kinds of stochastic dynamics can be easily derived [21].

The definition of the TPE also requires specification of the stable states A and B. Often, this can be done by demanding that one-dimensional order parameters $q_A(x)$ and $q_B(x)$ lie within appropriate limits. For instance, in the case of protein folding it may be possible to define the initial and final states through the number of native contacts. It is, however, important to realize that order parameters that are sufficient to specify the stable states are not necessarily suitable for describing the complete transition. (This distinguishes order parameters from reaction coordinates; see Sect. 7 for a discussion of this issue.) But while order parameters do not need to be good reaction coordinates, it is crucial that they clearly discriminate between the stable states in the sense that no point in A belongs to the basin of attraction of B and vice versa [124]. If the order parameters do not strictly discriminate between A and B, the path sampling procedure may yield pathways that are not truly reactive and may not sample the reactive trajectories at all. In complex systems the definition of suitable order parameters is not a trivial issue and may require some trial-and-error experimentation.

5.2 Sampling the Transition Path Ensemble

The central idea of TPS now is to generate reactive trajectories with a frequency proportional to their probability in the TPE of (50). The generated pathways can then be further analyzed to yield information on reaction rates and mechanisms. One way to sample the TPE is by a MC procedure. In this approach, which is analogous to the MC simulation of, say, a molecular liquid, a random walk through trajectory space is carried out in a way such that trajectories are sampled according to their statistical weight. Here, the basic MC step consists of generating a new trajectory $x^{(n)}(\mathcal{T})$, the so-called trial trajectory, from an old one $x^{(o)}(\mathcal{T})$ by some procedure that we will specify later. This newly generated trajectory is then accepted or rejected depending on how the statistical weight of the new trajectory in the TPE compares to that of the old one. In case of an acceptance, the new trajectory becomes the current one. If, on the other hand, the trial trajectory is rejected, the old trajectory remains the current one. Iterating this procedure yields trajectories distributed according to the TPE provided appropriate acceptance/rejection rules are used.

To derive an acceptance/rejection rule for pathways one can start from the detailed balance condition

$$\mathcal{P}_{AB}\left[x^{(o)}(\mathcal{T})\right]\pi\left[x^{(o)}(\mathcal{T})\rightarrow x^{(n)}(\mathcal{T})\right] =$$
$$\mathcal{P}_{AB}\left[x^{(n)}(\mathcal{T})\right]\pi\left[x^{(n)}(\mathcal{T})\rightarrow x^{(o)}(\mathcal{T})\right], \tag{63}$$

which guarantees that the right ensemble is generated by requiring that the move from an old path to a new path is exactly balanced by the reverse move from a new path to an old one. According to the two-step nature of the MC procedure, the transition probability $\pi[x^{(o)}(\mathcal{T})\rightarrow x^{(n)}(\mathcal{T})]$ to move from the old path $x^{(o)}(\mathcal{T})$ to the new path $x^{(n)}(\mathcal{T})$ is the product of the probability P_{gen} to generate the new path and the probability P_{acc} to accept it,

$$\pi\left[x^{(o)}(\mathcal{T})\rightarrow x^{(n)}(\mathcal{T})\right] = P_{gen}\left[x^{(o)}(\mathcal{T})\rightarrow x^{(n)}(\mathcal{T})\right]$$
$$\times P_{acc}\left[x^{(o)}(\mathcal{T})\rightarrow x^{(n)}(\mathcal{T})\right]. \tag{64}$$

Insertion of this form of the transition probability into the detailed balance condition provides a condition for the acceptance probability that can be satisfied with the so-called Metropolis rule [174]. The resulting acceptance probability is given by [124]:

$$P_{acc}\left[x^{(o)}(\mathcal{T})\rightarrow x^{(n)}(\mathcal{T})\right] = h_A\left[x_0^{(n)}\right]h_B\left[x_{\mathcal{T}}^{(n)}\right]$$
$$\times \min\left\{1, \frac{\mathcal{P}\left[x^{(n)}(\mathcal{T})\right]P_{gen}\left[x^{(n)}(\mathcal{T})\rightarrow x^{(o)}(\mathcal{T})\right]}{\mathcal{P}\left[x^{(o)}(\mathcal{T})\right]P_{gen}\left[x^{(o)}(\mathcal{T})\rightarrow x^{(n)}(\mathcal{T})\right]}\right\}. \tag{65}$$

It follows from this expression that only reactive trajectories, i.e., trajectories for which $h_A[x_0^{(n)}] = 1$ and $h_B[x_{\mathcal{T}}^{(n)}] = 1$, can have a non-vanishing probability to be accepted. Equation (65) provides a general expression from which the specific acceptance probability for a particular path generation procedure can be derived.

The efficiency of a TPS simulation, i.e., the rate at which trajectory space is sampled, crucially depends on how in detail new trajectories are generated from old ones. While various ways to do that have been proposed [164], the so-called shooting algorithm has proven particularly useful (see Fig. 8). Since it is generally

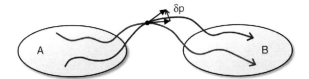

Fig. 8 In the shooting algorithm for deterministic dynamics a new path (*blue*) is generated from an old one (*red*) by first randomly selecting one point on the old path, the *shooting point*. Then, the particle momenta at that point are modified by addition of a small perturbation δp. From the point with perturbed momenta the equations of motion are integrated forward and backward to obtain a complete trajectory. For small perturbations, the new trajectory will be close to the old one near the shooting point but will then rapidly diverge from it due to the chaoticity of the underlying dynamics

applicable and can be used to illustrate the general path sampling MC procedure, we will briefly outline this method here. The basic idea of the shooting algorithm is to exploit the natural tendency of the dynamics to converge towards the stable states. In this procedure a new pathway is generated from an old one by picking a random time slice $x_{t'}^{(o)}$ of the old path. Then, $x_{t'}^{(o)}$ is modified, for instance by adding a small perturbation to the momenta, yielding $x_{t'}^{(n)}$. (For stochastic dynamics no perturbation is required, since the random noise will lead to different trajectories even if they start from the same phase space point.) Starting from this modified state, the equations of motion are integrated forward and backward to complete the new pathway. Since in the shooting algorithm new trajectories are generated according to the rules of the underlying dynamics, which we assume to be microscopically reversible here, most factors in (65) cancel and the acceptance probability is particularly simple,

$$P_{\text{acc}}\left[x^{(o)}(\mathcal{T}) \to x^{(n)}(\mathcal{T})\right] = h_A\left[x_0^{(n)}\right] h_B\left[x_{\mathcal{T}}^{(n)}\right] \min\left[1, \frac{\rho(x_{t'}^{(n)})}{\rho(x_{t'}^{(o)})}\right]. \qquad (66)$$

Here, $\rho(x)$ is the stationary distribution evaluated at x. For Newtonian dynamics and a microcanonical distribution of initial conditions the acceptance probability simplifies even further:

$$P_{\text{acc}}\left[x^{(o)}(\mathcal{T}) \to x^{(n)}(\mathcal{T})\right] = h_A\left[x_0^{(n)}\right] h_B\left[x_{\mathcal{T}}^{(n)}\right]. \qquad (67)$$

This expression implies that any new pathway that is reactive, i.e., that connects A and B, is accepted. In the shooting algorithm for deterministic trajectories the magnitude of the momentum perturbation can be used to control the average acceptance probability and hence to optimize the efficiency of the TPS simulation [175]. To enhance the ergodicity of a TPS simulation, it can be combined with PT [176] carried out at the path level [177]. To increase the efficiency of TPS simulations shooting moves can be complemented with so-called shifting moves and path reversal moves [165]. For diffusive barrier crossings it may be difficult to obtain a reasonable acceptance probability, because, due to the chaoticity of the underlying dynamics, a new trajectory generated by the shooting algorithm might be very different from the old one even for very small momentum displacements. Bolhuis has suggested an algorithm to circumvent this difficulty [178]. A further improvement relies on using linearized equations of motion for small displacements [179].

The rules described above provide a method to sample properly weighted pathways by carrying out an importance sampling simulation in path space. To start this procedure an initial pathway connecting A and B is required. One way to generate such an initial trajectory is to run a long MD trajectory and wait until the transition occurs spontaneously. In most situations, however, this may not be feasible. Often an initial pathway can be generated by driving the system from A to B artificially. For instance, biased trajectories obtained with steered MD can then be relaxed towards the right ensemble of pathways [180]. In the case of a pressure-induced solid–solid phase transition an initial transition pathway can be obtained by sufficiently over-pressurizing the system [181]. Such a trajectory does not carry a large statistical weight at less extreme conditions (its weight may even vanish), but can serve as a starting point for the simulation which then relaxes to more important parts of trajectory space. Similarly, a first folding trajectory of a protein may be obtained by letting the protein unfold at high temperature. Again, the initial trajectory obtained in this way is most likely not representative of the TPE at the temperature of interest, but nevertheless provides a starting point for the TPS procedure. In other cases, it may be possible to construct an initial reactive trajectory by hand [182]. Although such a trajectory may not even be a truly dynamical path, it often suffices to initiate the sampling procedure.

5.3 Kinetics from the Transition Path Ensemble

TPS harvests a large collection of trajectories connecting the initial to the final state. However, this ensemble itself does not contain enough information to compute the primary kinetic experimental observable, the rate constant. Nevertheless, the rate constant can be computed by an additional procedure from the correlation function introduced in Sect. 3 [165]

$$C(t) \equiv \frac{\langle h_A(x_0) h_B(x_t) \rangle}{\langle h_A(x_0) \rangle}, \tag{68}$$

with h_A and h_B defined by (51). Because of the separation of timescales, this population correlation function grows linearly in time, $C(t) \sim k_{AB} t$, for times $\tau_{mol} < t \ll \tau_{rxn}$. In that case, the time-dependent reaction rate

$$k_{AB}(t) = \dot{C}(t) \tag{69}$$

reaches a plateau for $\tau_{mol} < t \ll \tau_{rxn}$. To calculate the rate constant with TPS we express (68) as the ratio of two path ensemble averages,

$$C(t) = \frac{\int \mathcal{D}x(t)\mathcal{P}[x(t)]h_A(x_0)h_B(x_t)}{\int \mathcal{D}x(t)\mathcal{P}[x(t)]h_A(x_0)}, \tag{70}$$

where the integrals are over all possible paths of length t. The denominator does not depend on time and equals the equilibrium population in A, $\langle h_A \rangle$.

The next step is to define an order parameter $\lambda(x)$ which can be used to describe region $B \equiv \{x : \lambda_{min}^B < \lambda(x) < \lambda_{max}^B\}$ as well as the entire configuration space, $-\infty < \lambda(x) < \infty$, including A . Substitution of the indicator function $h_B(x)$ into (70) and change of the integration order leads to

$$C(t) = \frac{1}{\langle h_A \rangle} \int \mathcal{D}x(t)\mathcal{P}[x(t)]h_A(x_0) \int_{\lambda_{min}^B}^{\lambda_{max}^B} d\lambda\, \delta[\lambda - \lambda(x_t)] \tag{71}$$

$$= \int_{\lambda_{min}^B}^{\lambda_{max}^B} d\lambda\, \langle \delta[\lambda - \lambda(x_t)] \rangle_A \equiv \int_{\lambda_{min}^B}^{\lambda_{max}^B} d\lambda\, P_A(\lambda, t),$$

where $\langle \cdots \rangle_A$ denotes path averaging over all trajectories originating in A. The function $P_A(\lambda, t)$ is the probability that at time t a path has reached λ, provided it started in A. As the process of interest is rare, we are naturally dealing with low probabilities. A path sampling equivalent of the US technique [42] can solve this problem by dividing the λ-range into a number of windows, and computing the following probability for each window W_i defined by $\lambda_i^{min} < \lambda(x) < \lambda_i^{max}$:

$$P_{AW_i}(\lambda, t) = \frac{\int \mathcal{D}x(t)\mathcal{P}[x(t)]h_A(x_0)h_{W_i}(x_t)\delta[\lambda - \lambda(x_t)]}{\int \mathcal{D}x(t)\mathcal{P}[x(t)]h_A(x_0)h_{W_i}(x_t)}$$

$$= \langle \delta[\lambda - \lambda(x_t)] \rangle_{AW_i}. \tag{72}$$

Rematching the probabilities for all windows eventually yields $P_A(\lambda, t)$ and hence through (71) correlation function $C(t)$.

The combination of a path sampling simulation employing the shooting and shifting MC moves, with an US algorithm in which the final region is transformed continuously from spanning the entire phase space to only the final stable state of interest B, yields $P_A(\lambda)$, and, through (71), $C(t)$ and, hence, the rate constant [165].

While the US procedure could be repeated for every t to yield the full correlation function $C(t)$, a computationally more convenient approach exists [21, 175], which allows us to write the rate constant as

$$k(t) \equiv \dot{C}(t) = \frac{\langle \dot{h}_B(t) \rangle_{AB}}{\langle h_B(t') \rangle_{AB}} C(t'), \tag{73}$$

where the first factor can be computed in a path sampling simulation with a fixed length t. Knowledge of $C(t')$ at time $t' < t$, leads to $k(t)$ at all times t by multiplying by this factor. Hence, the computationally expensive US scheme only has to be carried out once for a time t' which can be much shorter than t. The estimate of the first factor can be improved by using a special indicator function that is unity whenever the path only visits B but does not necessarily end in it. We refer to [165] for more details on this algorithm. A method for calculating activation energies rather than full reaction rate constants has been presented in [183]. Applying a kind of TI [43] in the space of trajectories, this method can also be used to calculate reaction rate constants by starting from a state with known reaction rate constant and slowly transforming the path ensemble into the ensemble of interest.

5.4 Transition Interface Sampling

The method discussed in the previous section is not necessarily the most efficient way of calculating the kinetics in the TPS method. In particular, the requirement that the path length has to be fixed a priori, plagues efficient implementation of path sampling, even when applying the convenient factorization of (73) [184]. In contrast, the transition interface sampling (TIS) method was developed with a flexible path length in mind. This method defines a series of $n + 1$ multidimensional interfaces by means of a suitable order parameter λ_i (just like the windows in the previous section) and measures the effective positive flux through these interfaces (see Fig. 9). The heart of the method is an expression for the rate constant that is essentially a reinterpretation of the reactive flux,

$$k_{AB} = \Phi_{1,0} P_A(\lambda_n | \lambda_1). \tag{74}$$

The first term on the right-hand side, $\Phi_{1,0}$, is the effective positive flux of trajectories that leave the stable state A (the boundary of which is defined by λ_0) through the

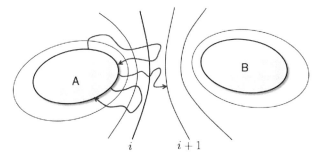

Fig. 9 In the TIS method, a series of non-intersecting interfaces between regions A and B are defined and the effective positive flux through these interfaces is measured. This is done in separate path sampling simulations for each pair of adjacent interfaces. In the path ensemble belonging to interfaces i and $i + 1$, only pathways that start in A, cross interface i and then either cross interface $i + 1$ or return to region A can have a non-vanishing weight (*red pathways*). Pathways that do not cross interface i are not part of this ensemble (*blue path*)

first interface λ_1. Here, "positive" means that only crossings in the forward direction towards B are counted and "effective" implies that recrossings are not counted before the trajectory goes back to A. The factor $\Phi_{1,0}$ is easily computed by conducting a MD simulation in state A and counting the number of first crossings of interface λ_1 after the trajectory has left A, per unit time. The second term $P_A(\lambda_n|\lambda_1)$ is the conditional probability that a trajectory coming from A and crossing λ_1 reaches λ_n, which defines the boundary of B. This so-called crossing probability is more difficult to compute, as the crossing is a rare event. However, this product can be decomposed in a product of conditional probabilities,

$$P_A(\lambda_n|\lambda_1) = \prod_{i=1}^{n-1} P_A(\lambda_{i+1}|\lambda_i). \tag{75}$$

Again $P_A(\lambda_j|\lambda_i)$ denotes the probability that a trajectory that leaves A, and crosses λ_i, will reach λ_j before returning to A. Each of the factors in this product can be computed in a path sampling simulation. This path sampling is done using the TIS path ensemble, that, in contrast to the regular TPE, is defined as the collection of all paths that leave A, cross λ_i, and either reach λ_j or return to A as illustrated in Fig. 9. The characteristic function $\tilde{h}_{ij}[x;\mathcal{T}]$ is unity for such paths and zero otherwise. The TIS path sampling scheme consists of choosing a random slice on an existing path that leaves A and crosses λ_i, reaches λ_j or returns to A. From this slice a new path is generated using the shooting algorithm, just as in the method described in the previous sections. The main difference is that the integration is stopped when the trajectory reaches A, or interface λ_j. The new path is accepted when it is part of the TIS ensemble, i.e., if it leaves A, crosses λ_i, and either reaches λ_j or returns to A. The acceptance criterion is then

$$P_{\text{acc}}\left[x^{(o)}(\mathcal{T}^{(o)}) \to x^{(n)}(\mathcal{T}^{(n)})\right] = \tilde{h}_{i,i+1}\left[x^{(n)}(\mathcal{T}^{(n)})\right]$$

$$\times \min\left[1, \frac{\mathcal{T}^{(o)}}{\mathcal{T}^{(n)}} \frac{\rho(x_{t''}^{(n)})}{\rho(x_{t'}^{(o)})}\right]. \tag{76}$$

Here, the factors $\mathcal{T}$ in the min-function account for the generating probability of choosing the shooting slice. Note that t' and t'' denote the same slice, but are shifted due to the change in path length. In addition to TIS shooting moves, sampling can be enhanced by reversal of the pathways that start and end in A. While not creating a new path, reversals allow for better exploration of the path space. The shifting move is not required in TIS, as the paths end precisely at the boundary of A and the interface $i+1$. Note that the TIS path sampling has to be performed n times to compute the rate constant. The crossing probability $P_A(\lambda|\lambda_1)$ as a function of λ will monotonously decrease until a plateau is reached in λ, which is equal to the desired crossing probability. This feature is conceptually different from the plateau in the time correlation function in previous sections. Nevertheless, the resulting value of the rate constant is independent of the path sampling method used and only the efficiency and convergence properties might be affected. The flexible path length can also be implemented in the regular TPS algorithm, only sampling A–B trajectories [185].

Van Erp and Bolhuis proposed several enhancements of TIS in [138], notably a configurational bias method to prune unsuccessful pathways.

5.5 Partial Path Sampling

The transition path simulation of processes involving diffusive barrier crossings can require very long trajectories. In some cases, the efficiency of the simulation can be dramatically enhanced by making use of loss of correlation along such long pathways. The idea of the loss of correlation led to the partial path TIS (PPTIS) method [186]. Here, instead of creating pathways that leave A, cross an interface, and then return to A or continue to the next interface, one considers short trajectories (partial paths) that only span one or two interfaces. The framework of PPTIS is that of TIS, except that the crossing probability now can depend on different starting and ending interfaces. In particular, the PPTIS defines the single interface crossing probabilities $p_i^{\pm}, p_i^{\mp}, p_i^{=}, p_i^{\ddagger}$. These quantities denote the different probabilities that a path that crosses i starts or ends at the $i-1$ or $i+1$ interface. These probabilities can be computed by a path sampling algorithm using only very short paths. The integration can be stopped at the $i-1$ or $i+1$ interface. Both the forward and backward rate constants can be determined:

$$k_{AB} = \frac{\langle \phi_{1,0} \rangle}{\langle h_{\mathcal{A}} \rangle} P_n^+, \qquad k_{BA} = \frac{\langle \phi_{n-1,n} \rangle}{\langle h_{\mathcal{B}} \rangle} P_0^- . \tag{77}$$

Here P_n^+ and P_n^- denote the long-distance crossing probabilities. For instance, P_i^+ is the probability that a trajectory crosses i while coming from A directly (recall that $\lambda_0 = \lambda_A$, and $\lambda_n = \lambda_B$). P_i^- is defined likewise for the reverse direction. Now, it is possible to construct these long-distance crossing probabilities from single-interface crossing probabilities by the following recursive relation [186]

$$P_j^+ \approx \frac{p_{j-1}^{\pm} P_{j-1}^+}{p_{j-1}^{\pm} + p_{j-1}^{=} P_{j-1}^-}, \qquad P_j^- \approx \frac{p_{j-1}^{\mp} P_{j-1}^-}{p_{j-1}^{\pm} + p_{j-1}^{=} P_{j-1}^-}. \tag{78}$$

Here the big advantage with respect to TIS is that the paths become much shorter, and hence a PPTIS approach is more efficient.

The assumption made in PPTIS is that there is sufficient memory loss between the interfaces to justify the shorter paths. Hence, the interfaces should be chosen sufficiently far apart. Moroni et al. devised special memory loss functions to test the PPTIS assumption [186]. Van Erp and Bolhuis proposed in [138] a powerful combination of RE and (PP)TIS. Such a combination was shown to enhance the sampling efficiency dramatically in [187, 188]. The (PP)TIS can also be used for a simultaneous calculation of the reaction rate constant and the free energy along a selected reaction coordinate [189].

5.6 Forward Flux Sampling

The forward flux sampling method (FFS) was conceived by Allen, Warren and ten Wolde [190–192] to deal with stochastic non-equilibrium systems in which the phase space distribution is not a priori known. Inspired by the TIS method, FFS employs the notion of a set of interfaces along a reaction coordinate (see Fig. 9), and even uses the same central reaction rate expression,

$$k_{AB} = \Phi_{1,0} P_A(\lambda_n|\lambda_1). \tag{79}$$

As in (PP)TIS, the flux factor can easily be computed by a straightforward MD simulation. The main difference between FFS and TIS is the way that the crossing probability is computed. Path sampling is based on microscopic reversibility, and time reversal of trajectories, but for systems with inherent non-reversible dynamics, reversal of trajectories is not an option. The prime example of such a system is an irreversible chemical network in which there are sinks and sources. The change in concentrations can be computed by stochastic dynamics in the forward direction, but not backward. The solution of FFS is to do away with the backward shooting part of TPS and only shoot forward from previous paths. The algorithm is bootstrapped from a regular dynamical simulation. Whenever the first interface λ_1 is crossed, the crossing point is stored in memory, and the trajectory is halted and reinitialized in state A. The resulting ensemble of crossing points is then used as initial points for the computation of the crossing probability of the second interface λ_2. This is done by choosing, from the ensemble of initial crossing points, a random configuration and integrate only forward until the next interface is crossed, or the path returns to the initial state A. Because of the stochasticity of the dynamics, paths will diverge even when starting from the same initial point. The shooting continues until there are enough crossing points in the ensemble of the second interface. This procedure is repeated for each interface until λ_n is reached. The forward flux approach not only yields the rate constant, but also the complete transition pathways from A to B can be reconstructed by gluing the successful shots together.

An advantage of FFS is that it does not require backwards pathways. Moreover, because it is not a MC Markov chain, it does not suffer from decorrelation times that plague Markov chains in general. In addition, FFS is not limited to non-equilibrium dynamics. It is also applicable to regular KMC and even (stochastic) MD.

Nevertheless, FFS has several drawbacks that are worth mentioning. The first is that the accuracy of the sampling very much depends on the quality of the first interface ensemble. Once the ensemble is under-sampled, errors propagate through the next interfaces. This does not happen in a MC method where each individual interface ensemble eventually converges. The other drawback is that the FFS does not allow pathways to relax their initial part. This follows naturally because the part of the path before the interface of interest is always fixed. Because of this effect, FFS is more dependent on a good reaction coordinate than for instance TIS, and certainly than TPS. Despite these drawbacks FFS is a powerful method, that has been used in many applications such as nucleation [193], the folding of a lattice protein [194], and chemical networks [190].

Allen and coworkers proposed several extensions of the FFS method in [192].

5.7 Milestoning

While not a path sampling simulation method in the strict sense, the milestoning method of Elber and coworkers [195, 196] has enough similarity with PPTIS and FFS to warrant a discussion in this section of the review. The Milestoning method starts with a number of hyper-surfaces (very similar to the TIS interfaces) for which a constraint equilibrium ensemble is prepared, for instance, according to the Boltzmann distribution. Starting from this ensemble, trajectories are initiated that are continued until they reach a neighboring hyper-surface. The distribution of times (path length) $K_s(t)$ of these trajectories that lead to the neighboring hyper-surface yields the entire kinetics of the system through the integral equations

$$P_s(t) = \int_0^t \left[1 - \int_0^{t-t'} K_s(\tau) d\tau \right] Q_s(t'), \tag{80}$$

$$Q_s(t) = \eta_s \delta(t) + \int_0^t \left[K_{s+1}^-(t-t') Q_{s+1}(t') + K_{s-1}^-(t-t') Q_{s-1}(t') \right] dt', \tag{81}$$

where n_s is the initial milestone distribution. Here, $P_s(t)$ denotes the probability that s is the last crossed milestone. $Q(t)$ denotes the probability to make a transition to milestone s at time t. The first equation states that $P_s(t)$ is equal to the probability to have come there at an earlier time t' and have not yet left. The second equation gives the probability Q_s to make a transition to a milestone s as a sum of the initial distribution and the probability to first reach a neighboring milestone and then hop to s. Together these two equations determine the $P_s(t)$. From this both the kinetics and the free energy can be determined.

The milestoning method provides a non-Markovian model for the kinetics because it contains the explicit time dependence of the distribution, which does not have to be exponential [196]. Milestoning also does not depend on the separation of timescales that most of the other methods take as a starting point.

The milestoning method can be extended to more than one order parameter. Elber and coworkers have applied the method to biomolecular systems [196, 197]. For a conformational transition of alanine dipeptide, they found an order of magnitude efficiency enhancement with respect to straightforward MD.

5.8 Transition Path Sampling Applications

To date, TPS has been applied to processes in many fields ranging from physics and materials science to chemistry and biology. One class of problems that has been successfully addressed with TPS are first-order phase transitions in condensed materials. Above the spinodal, such transitions occur via the formation of a critical nucleus of the stable phase in the metastable phase. Because of the free energetic cost associated with the creation of an interface between the two phases, this process involves the crossing of a free energy barrier. For this reason, nucleation is a rare event and requires special computational techniques [198]. Nucleation processes that have been studied with TPS include magnetization reversal in the Ising-model [199], pressure-induced phase transitions in semiconductor nanoclusters [182], the freezing of Lennard-Jonesium [200], phase separation and crystallization from the melt [201], the solid–solid transition of terephthalic acid [202], the liquid–vapor transition of methane [203], the wurtzite to rocksalt transition in bulk CdSe [204], the boiling of water [205], pressure-induced transitions of alkali halides [206–211], and crystallization from solution [212].

Chemical processes often involve rare events because of high energy barriers that have to be crossed during reactions or entropic effects that are due to complex solvent rearrangements. Chemical processes addressed with TPS methodologies include proton transfer in the water trimer [128], autoionization in liquid water [7], hydrated proton transfer in water [213], the dissociation of acetic acid [214], C–C bond formation in the methanol coupling reaction in chabazite [215], ligand exchange at a Cr metal center [216], acid-catalyzed peptide hydrolysis [217], incorporation of Helium into C-60 [218], the dissociation of hydrogen peroxide with iron(II) in aqueous solution [219], the Cl^- + $ClCH_2CN$ S_N2 reaction [220], kinetic pathways of ion dissociation in water [25, 221], the isomerization and melting of water clusters [222, 223], dynamics of hydrogen bonds in water [224], solvation of NaCl [225, 226], cavitation between hydrophobic surfaces [227], diffusion of isobutane in silicalite [228], micelle fusion and fission [229], and the trans-gauche transition in liquid n-butane [230].

The application of TPS to biological processes has been recently reviewed in [169].[8] Biological processes investigated with TPS comprise the isomerization of alanine dipeptide [26, 28, 233], the binding and unbinding of DNA base pairs [234], the chorismate-mutase-catalyzed conversion of chorismate into prephenate (enzyme-catalyzed reaction) [235], the collapse of a hydrophobic homopolymer in solvent [236], the folding of the GB1 beta hairpin [237, 238], the folding of the Trp-cage mini-protein [185], the DNA repair process by polymerase [231, 239, 240], the enzymatic reaction of the lactate dehydrogenase [241], the flip-flop of lipids in

---

[8] On page 311 of [169] we incorrectly stated that the BOLAS method put forward by Radhakrishnan and Schlick [231] is slightly biased. The BOLAS method, described in more detail in a subsequent publication [232], is correct and can be used to study free energy barriers for complex (biological) processes.

membranes [242, 243], the activation mechanism of a signaling protein [244], and closing pathways for DNA polymerase β [245].

TPS has also been used to address the glass transition [246], non-equilibrium dynamics [247, 248], the sampling of trajectories with rare work values in the context of Jarzynski's non-equilibrium theorem [249–254], as well as the calculation of entropy flow distribution functions in driven stochastic systems [255, 256].

6 Discrete Path Sampling

6.1 Discrete Path Sampling Theory and Algorithm

Wales and coworkers have developed a discrete path sampling (DPS) procedure to compute rate constants and mechanism in complex systems [257, 258]. Although this procedure is slightly different in spirit to the path sampling schemes mentioned above, it is worthwhile to discuss this approach. The systems under consideration are transitions that occur via a network of intermediate metastable states. Known examples are the rearrangement of crystal structures and the rearrangement of finite (molecular) clusters, but also conformational changes in biomolecules. DPS is essentially a way to sample Markovian state models efficiently. The advantage is that one does not need to know all minima and stationary points of the system in advance, as is the case with KMC or master equation approaches. Rather, one creates on-the-fly a database of minima and transition states representing the fastest overall transition pathways.

In the DPS method, one starts with the master equation

$$\frac{dp_\alpha(t)}{dt} = \sum_{\beta \neq \alpha} k_{\beta\alpha} p_\beta(t) - k_{\alpha\beta} p_\alpha(t), \tag{82}$$

where p_α denotes the population probability in state α and $k_{\alpha\beta}$ is the rate constant from state α to β. If all metastable states and their respective rate constants are known, then the solution of the master equation is straightforward. In practice, however, this would be a daunting task for most complex rearrangements. The DPS approach is an attempt to circumvent such an exhaustive computation by obtaining the most relevant pathways contributing the most to the overall rate.

When the network consists of metastable states such that the system stays for a long time in either globally stable states A or B, and only rarely crosses the intermediate states $i \notin A \cup B$ then the master equation can be rewritten in terms of overall rate constants using two approximations. The first is the approximation of local equilibrium in A and B, i.e., the probability $p_a(t)$ to be in a minimum a belonging to A is proportional to the overall probability to be in A, P_A, according to $p_a(t) = P_A(t)p_a^{eq}/P_A^{eq}$, where the superscript "eq" denotes the equilibrium value. The second approximation is the steady state assumption for all intermediate states

i, $\mathrm{d}p_i(t)/\mathrm{d}t = 0$. The overall master equation thus reads

$$\frac{\mathrm{d}P_A(t)}{\mathrm{d}t} = -k_{AB}P_A(t) + k_{BA}P_B(t), \tag{83}$$

$$\frac{\mathrm{d}P_B(t)}{\mathrm{d}t} = +k_{AB}P_A(t) - k_{BA}P_B(t), \tag{84}$$

where

$$k_{AB} = \frac{1}{P_A^{\mathrm{eq}}} \sum_{a,i_1,i_2,\ldots,i_n,b} \frac{p_a^{\mathrm{eq}}k_{ai_1}k_{i_1i_2}\cdots k_{i_nb}}{\sum_{j_1}k_{j_1i_1}\sum_{j_2}k_{j_2i_2}\cdots\sum_{j_n}k_{j_ni_n}}, \tag{85}$$

$$k_{BA} = \frac{1}{P_B^{\mathrm{eq}}} \sum_{b,i_n,\ldots,i_2,i_1,a} \frac{p_b^{\mathrm{eq}}k_{bi_n}\cdots k_{i_2i_1}k_{i_1a}}{\sum_{j_1}k_{j_1i_1}\sum_{j_2}k_{j_2i_2}\cdots\sum_{j_n}k_{j_ni_n}}. \tag{86}$$

Here, the sum is over all possible paths from a minimum $a \in A$ via an arbitrary number n of intermediates $i \notin A \cup B$ to $b \in B$. The rate constants k_{ij} for each transition between intermediate i and j are computed from harmonic TST (see Sect. 3.3).

In practice these sums can be infinite, due to recrossings, and therefore one introduces a method to compute the contribution due to recrossings based on the shortest paths between A and B, by defining a propagation matrix A:

$$A = \begin{Bmatrix} 0 & k_{i_2i_1}/\sum_{\gamma}k_{i_1\gamma} & 0 & \cdots & 0 \\ k_{i_1i_2}/\sum_{\gamma}k_{i_2\gamma} & 0 & k_{i_3i_2}/\sum_{\gamma}k_{i_2\gamma} & \cdots & 0 \\ 0 & k_{i_2i_3}/\sum_{\gamma}k_{i_3\gamma} & 0 & \cdots & 0 \\ \cdots & \cdots & \cdots & & \cdots \end{Bmatrix}, \tag{87}$$

where the sums are over all neighbor states γ.

The contribution to the rate constant of a $a - b$ or $b - a$ path, respectively, is then

$$k_{ab} = \frac{p_a^{\mathrm{eq}}}{P_A^{\mathrm{eq}}}k_{i_nb}\frac{k_{ai_1}}{\sum_{\gamma}k_{i_1\gamma}}\sum_{p=n-1}^{\infty}[A^p]_{n1}, \tag{88}$$

$$k_{ba} = \frac{p_b^{\mathrm{eq}}}{P_B^{\mathrm{eq}}}k_{i_1a}\frac{k_{bi_n}}{\sum_{\gamma}k_{i_n\gamma}}\sum_{p=n-1}^{\infty}[A^p]_{1n}, \tag{89}$$

where the subscript refers to the corresponding matrix element. The corresponding total rate constants k_{AB} and k_{BA} are given by

$$k_{AB} = \sum_{a,i_1,i_2,\ldots,i_n,b} k_{ab}, \tag{90}$$

$$k_{BA} = \sum_{b,i_n,\ldots,i_2,i_1,a} k_{ba}. \tag{91}$$

The DPS algorithm consists of starting from an initial path $\{a,i_1,\ldots,i_n,b\}$ and perturbing this path by replacing a random intermediate or including a new intermediate in the path. When the $a - b$ rate constant is found to become larger, then the trial path replaces the current path. The sampling is finished when the fastest possible path has been found. The overall kinetics is assumed to be dominated by the fastest pathways.

The DPS method can be improved by the application of a graph transformation [259], in order to decrease the number of minima in the database, while keeping the properties of interest unchanged.

6.2 DPS Applications

Wales applied this scheme to several systems including the two-dimensional seven particle LJ cluster (heptamer), the $(H_2O)_8$ water cluster, and the LJ_{38} cluster [258]. In a later paper, Evans and Wales applied DPS to the folding of the GB1 hairpin in implicit solvent [260]. Employing the DPS scheme, they created a database of several tens of thousand of minima and transitions states. A KMC simulation on this database yielded an estimated folding time of around 30–90 μs, about ten times slower than the experimental one, which is reasonable considering the approximation made in the implicit force field and the harmonic approximation. Using graph theoretical algorithms they found that the fastest path only contributed with a folding rate $k_{aB} = 10^{-48}\,\mathrm{s}^{-1}$, about 50 orders of magnitude slower than the KMC value or the experimental rate ($\approx 10^6\,\mathrm{s}^{-1}$). This indicates that for this system the number of pathways that should be included into the ensemble is truly enormous.

7 Reaction Coordinate and Committor

A collection of transition pathways in full atomistic detail, for instance harvested with the TPS methods or from a long MD trajectory, does not directly result in a detailed understanding of the underlying mechanism. The situation is similar to that encountered when performing a straightforward MD simulation of a complex molecular system, say, a protein in aqueous solution: a detailed trajectory, stored on a computer in the form of a long list of particle positions and momenta at consecutive times, does not automatically generate understanding of the simulated system. Only further statistical analysis of the trajectory, perhaps guided by intuition, yields useful information and helps to identify those variables that capture the relevant physical features. Building on this insight, one can then describe the essential physics in terms of low-dimensional models in which all irrelevant degrees of freedom have been removed. Similarly, only further statistical analysis, carried out on a given set of pathways or on-the-fly as the pathways are generated, helps to extract a description of the mechanism in terms of a few important variables, or, ideally, to find a good reaction coordinate.

The reaction coordinate is a function $q(r)$, usually defined in configuration space, whose value is presumed to provide a measure for the progress of a reaction. For chemical reactions one may, for instance, expect a particular bond length or bond angle to serve as a reaction coordinate. For a folding protein, the number of native contacts may seem to lend itself as a reaction coordinate, and for a dissociating ion pair the interionic distance may be assumed to quantify how far the dissociation has proceeded. If judiciously chosen, such parameters may indeed change in a continuous manner from one value characteristic for the state A to another value typical for state B as the reaction occurs. The interionic distance certainly increases as the ions dissociate, and native contacts form as the protein folds. But does that imply that these parameters are "good" reaction coordinates? And what does "good reaction coordinate" mean and how can we distinguish it from a "poor reaction coordinate"?

It is important to realize that in general the reaction coordinate $q(r)$, like the collective variables we use for free energy calculations, is a function we define with some arbitrariness on the basis of what we already know about the process we would like to study (we will discuss procedures that can be employed to facilitate the search for a good reaction coordinate later). Our particular choice of $q(r)$ may or may not be suitable to describe the reaction of interest. From a good reaction coordinate we expect to be able to tell how far a reaction has proceeded and what will most likely happen next.[9] The reaction coordinate $q(r)$ should, for instance, tell us whether a particular configuration r is a transition state, i.e., a configuration from which both states A and B are equally accessible. By looking at the reaction coordinate $q(r)$ only, we should also be able to tell whether a reaction has just started or is about to be completed.

In the rare events context one usually distinguishes between a reaction coordinate and an order parameter. While the former is required to be a dynamically relevant measure for the complete progress of the reaction from start to finish, the latter is a variable that permits to discriminate between the stable states A and B but is not necessarily suitable for describing the course of the reaction. A good reaction coordinate can serve also as a good order parameter, but the inverse is not necessarily true.

We can make the concept of the quality of a reaction coordinate $q(r)$ more precise by considering the so-called commitment probability, or committor. The committor $p_B(r)$ is defined as the probability that a trajectory started at configuration r with random momenta reaches state B before it reaches state A (see Fig. 10). (The commitment probability for state A is defined analogously.) The commitment probability was introduced as splitting probability already by Onsager, who used this concept to analyze ion pair recombination [262]. It has proven very useful in theoretical studies of protein folding, where the committor is known as p_{fold} [263], and even in experimental work on liquid–solid nucleation [264]. Calculation of the probability $p_B(r)$ involves a Maxwell–Boltzmann average over momentum space

---

[9] The ability to predict the likely fate of a trajectory passing through a configuration r solely from the value of the reaction coordinate $q(r)$ at that configuration implies that the dynamics of the system projected onto the reaction coordinate [173] is, at least approximately, Markovian [261].

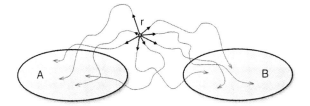

Fig. 10 The committor $p_B(r)$ for a particular configuration r measures the probability of a trajectory started at r to relax into region B. Numerically, the committor can be calculated by starting N trajectories from r, initialized with random momenta, and counting how many of them reach B rather than A

and, in the case of stochastic dynamics, also an average of noise histories.[10] The committor is a statistical measure for how committed a particular configuration is to state B. Configurations in or near region A will most likely have a very small committor, $p_B \approx 0$, while configurations in or near B will have a committor close to unity, $p_B \approx 1$. It is now natural to define the transition states as those intermediate configurations that have equal probability to reach A and B, or, in other words, are equally committed to either side, $p_A(r) = p_B(r) = 1/2$. The idea of using the committor to identify transition states goes back at least to Ryter [265, 266], and was used by several authors in the theory of activated stochastic processes [267–269].

How is the committor now related to the reaction coordinate? As mentioned above, one may expect a good reaction coordinate $q(r)$ to provide sufficient information to predict the likely fate of a trajectory passing through configuration r. But this is exactly what the committor $p_B(r)$ does. By looking solely at the committor we can tell how far a reaction has proceeded and what is likely to happen next. Committor values of 0 and 1 correspond to configurations that firmly belong to A and B, respectively, while a committor value of 1/2 implies that the system is at a transition state from which it can access both stable states with equal probability. Thus, in this almost tautological sense, the committor $p_B(r)$ itself is the perfect reaction coordinate [28, 31, 261].[11] This property of the committor also provides a criterion

---

[10] It is worth noting that different definitions of the committor exist. If the full microscopic description of the system includes momenta, the committor can be considered for a point in configuration space or for a point in phase space as is done in transition path theory (TPT) [31]. In the latter case, the average extends only over noise histories for stochastic dynamics and for deterministic dynamics the committor can only take the values 0 or 1. Furthermore, committor definitions can differ in whether one requires the trajectories to "reach" B before A or to "relax" into B rather than A. The latter definition is supposed to take into account correlated recrossing events into and out of B. Particularly for reactions taking place in the so-called energy diffusion regime, where the rate of energy dissipation is the main factor determining the kinetics, this definition seems more appropriate. Note, however, that such a committor definition is not suitable as a basis for the TPT discussed in Sect. 9, because some of the properties of the committor on which this theory is based are lost.

[11] It has been shown that for diffusive barrier crossing under certain not unduly restrictive conditions the reaction coordinate that is optimum in the TST-sense is orthogonal to the committor-1/2 surface [270].

for distinguishing a "good" reaction coordinate from a "poor" one. From any other "good" reaction coordinate $q(r)$ we require that it parametrizes the committor. In other words, the value of the reaction coordinate at configuration r needs to determine, at least to a good approximation, the value of the committor at that point, $p_B(r) = p_B[q(r)]$. For a "poor" reaction coordinate, on the other hand, there is no one-to-one relation to the committor. But while the committor is the ideal reaction coordinate in the sense explained above, it is very unspecific and does not directly lead to physical insight, i.e., it does not automatically yield a low-dimensional description of the system in terms of a few specific collective variables that distill the essential physics in a transparent way and can be probed or controlled in experiments or simulations. In the following section we will discuss several approaches for finding such variables.

8   Finding the Mechanism

Watching transition trajectories as molecular movies rendered on a computer is often a fascinating but at the same time sobering experience. While the suggestive images can provide useful insights and stimulate our imagination, important variables can remain elusive. For such complex reactions the committor can serve as a guide to identify relevant collective variables and test proposed reaction coordinates. In this section we briefly review several computational tools and approaches based on this concept.

8.1   Transition State Ensemble

Comparison of configurations with different committor values often yields valuable information on the reaction mechanism. In particular, it can be helpful to examine the properties of the transition state ensemble (TSE) which consists of the points where transition pathways pierce the isocommittor-1/2 surface, i.e., the surface defined by $p_B(r) = 1/2$. Since transition pathways cross the isocommittor-1/2 surface at different points that are not uniformly distributed, the TSE introduces a statistical weight on this surface. For diffusive dynamics (and/or a committor defined in phase space) the TSE is simply the equilibrium ensemble confined to the isocommittor-1/2 surface [31]. For non-diffusive dynamics and for a committor defined in configuration space, however, these two ensembles may differ. In a TPS simulation, the TSE can be determined by calculating the committor at regularly spaced points collected from transition pathways. Since along each transition pathway the committor grows continuously from $p_B = 0$ in region A to $p_B = 1$ in region B, the committor must be 1/2 (or close to it) for one or more intermediate configurations. These configurations are samples of the TSE. Analysis of the transition state ensemble has been proven useful in the investigation of various complex processes, including ion pair separation in water [25], biomolecular isomerization [26], and the liquid–solid transition [200]. Further information can be gained by analyzing the local flux through isocommittor surfaces [271].

8.2 Committor Distributions

As discussed above, a good reaction coordinate needs to parametrize the committor, i.e., configurations with a particular value of the reaction coordinate should all have the same committor. Therefore, isosurfaces of the reaction coordinate defined by $q(r) = $ constant should coincide, at least where they are mostly populated, with the corresponding isocommittor surfaces. This is something that can be easily tested by determining the probability distribution $P(p_B)$ of the committor for equilibrium-weighted configurations with a particular value q^* of the reaction coordinate [25]:

$$P(p_B) = \langle \delta[p_B - p_B(x)] \rangle_{q(r)=q^*}, \tag{92}$$

where $\langle \cdots \rangle_{q(r)=q^*}$ denotes an equilibrium average restricted to $q(r) = q^*$. If the reaction coordinate is sufficient to specify the value of the committor, the committor distribution $P(p_B)$ will be sharply peaked around the corresponding committor value, $p_B(q^*)$. For a poor reaction coordinate on the other hand, the committor distribution $P(p_B)$ will not be unimodal as configurations with the same value of the reaction coordinate can have different committors.

For a proposed reaction coordinate $q(r)$ it is instructive to determine the committor distribution $P(p_B)$ for the reaction coordinate constrained at the maximum q^* of the free energy barrier $F(q)$ that separates the stable states from each other (of course, this presupposes that the free energy as a function of q has such a barrier). If $q(r)$ is a good reaction coordinate, all configurations with $q(r) = q^*$ are transition states and the distribution $P(p_B)$ is concentrated around $p_B = 1/2$. A distribution $P(p_B)$ that does not have a single peak at $p_B = 1/2$ indicates that degrees of freedom other than $q(r)$ are necessary to specify transition states and quantify the reaction progress. In this case, different scenarios are possible [165]. If most configurations of the $q(r) = q^*$ ensemble belong to A and/or B, the committor distribution will be peaked at 0 and/or 1. If diffusive barrier crossing occurs at $q(r) = q^*$ but in a direction orthogonal to the $q(r) = q^*$ surface, the committor distribution is uniform in the interval from 0 to 1 [165].

To assess the quality of a proposed reaction coordinate $q(r)$ in practice, one has first to decide on the particular value q^* of the reaction coordinate at which the committor analysis should be carried out. This can be done on the basis of the free energy profile $F(q)$ determined using standard methods such as TI or umbrella sampling. Then, one generates a sample of equilibrium configurations with the reaction coordinate fixed at q^*, for instance by constrained MD or MC simulation. For each configuration the committor is calculated by "shooting off" a number of trajectories and determining the fraction of trajectories that relax into the final state. The committor values calculated in this way are histogrammed yielding an estimate of the committor distribution $P(p_B)$. A detailed statistical analysis of the computation of committor histograms has been carried out by Peters [272]. Initially introduced to study ionic dissociation in water [25], committor distribution analysis was subsequently applied to elucidate the mechanism of various complex reactions [26, 28, 29, 199, 200, 202, 273].

8.3 Bayesian Path Statistics

An alternative definition of transition states was recently suggested by Hummer [27,261] on the basis of a Bayesian relation between the TPE and the equilibrium ensemble. Since in the TPE pathways are constrained to connect A and B, the density distribution of microscopic states $P(x|TP)$ visited along transition pathways differs from the equilibrium distribution $\rho(x)$.[12] Using these two densities, the probability $P(TP|x)$ that a trajectory going through x is a transition path can be expressed as

$$P(TP|x) = \frac{P(x|TP)P(TP)}{\rho(x)}. \tag{93}$$

Here, the normalizing factor $P(TP) = \int dx \rho(x) P(TP|x)$ is the overall likelihood to be on a transition path averaged over all microscopic states x and it equals the fraction of time spent on transition pathways on a long equilibrium trajectory. The conditional probability $P(TP|x)$ is large at points x that are common to many transition pathways but are rarely visited in equilibrium. Therefore, it makes sense to identify transition states with those points at which $P(TP|x)$ is a maximum, i.e., with the points that have the largest probability that trajectories passing through them are reactive [27].

The probability $P(TP|x)$ is also simply related to the committor functions of the stable states, $P(TP|x) = p_A(\bar{x})p_B(x) + p_A(x)p_B(\bar{x})$, where $\bar{x} = \{r, -p\}$ results from $x = \{r, p\}$ by inversion of the momenta.[13] For diffusive dynamics, x consists of the positions r only and $P(TP|x) = 2p_B(x)[1 - p_B(x)]$. It follows that in this case the points r that maximize $P(TP|r)$ lie on the surface defined by $p_B = 1/2$ (at its maximum, $p(TP|r) = 1/2$). Thus, for diffusive dynamics the transition state definitions via $p_B(r)$ and $P(TP|r)$ are equivalent. In general, however, the points x maximizing $P(TP|x)$ are isolated and do not form a dividing surface. Using the probability density $P(x|TP)$, the TSE defined previously can be expressed as

$$\rho_{TSE}(x) = \delta[p_B(x) - 1/2] p(x|TP)/Z_{TSE}, \tag{94}$$

---

[12] In this context, a transition pathway is defined to consist only of the trajectory segment between A and B excluding points in A and B.

[13] Here, the committor is defined for a point x in phase space by averaging over noise histories but not over momenta. For Newtonian dynamics, the committor defined in this way is either 0 or 1.

where Z_{TSE} normalizes the distribution. It then follows from (93) that, as stated above, for a committor defined in phase space the TSE is equal to the equilibrium ensemble restricted to the isocommittor-1/2 surface [31].

While the Bayesian relation of (93) is correct in principle, the high dimensionality of phase space makes its application difficult. In practice, its generalization for projected dynamics is more useful, particularly for testing reaction coordinate candidates [27]. For a given reaction coordinate $q(x)$, the probability $P(\text{TP}|q)$ is defined as an average of $P(\text{TP}|x)$ over the equilibrium ensemble constrained at $q(x) = q$,

$$P(\text{TP}|q) = \frac{\int dx \rho(x) \delta[q - q(x)] P(\text{TP}|x)}{\int dx \rho(x) \delta[q - q(x)]}. \tag{95}$$

This probability is related to the density of q in the TPE and in the equilibrium ensemble by

$$P(\text{TP}|q) = \frac{P(q|\text{TP}) P(\text{TP})}{P(q)}. \tag{96}$$

The densities $P(q|\text{TP})$ and $P(q) \propto \exp[-\beta F(q)]$ can be obtained from a transition path simulation and an equilibrium free energy calculation, respectively. Equation (96) also provides the basis for a procedure to calculate reaction rate constants [27].

The probability $P(\text{TP}|q)$ calculated from (96) can be used to assess the quality of reaction coordinates. For a good reaction coordinate, all transition states, i.e., the states x with large probability $P(\text{TP}|x)$, should correspond to approximately the same value of the reaction coordinate. Hence, the probability $P(\text{TP}|q)$ should be a sharply peaked function with a maximum at the transition state value of q. For a poor reaction coordinate, on the other hand, $P(\text{TP}|q)$ is expected to be rather featureless according to (95), as in this case no strong correlation between the value of $q(x)$ and the probability $P(\text{TP}|x)$ exists. Best and Hummer have successfully used this approach to test reaction coordinates for the folding of a simple three-helix bundle protein and the collective dipole flip of ordered one-dimensional chains of hydrogen-bonded water molecules in narrow carbon nanotubes [261].

8.4 Genetic Neural Networks

Ma and Dinner recently used genetic neural networks (GNN) [274, 275] to automatically screen large sets of candidates for the reaction coordinate and identify the few collective variables which best parametrize the committor [28]. The method requires generation of a database computed from many different configurations of the system. Each entry of the database consists of the committor p_B and a possibly long list of collective variables all calculated for the same configuration of the system. To avoid a distorting bias it is important that the committor values included in the database are approximately uniformly distributed in the range from 0 to 1. A practical way to collect configurations for the database is by taking them from transition pathways, for instance harvested with TPS. The database is then divided into a training set and a test set. After optimizing the weights of the neural network on the training set for a given combination of a few collective variables, the quality of the fit is assessed on the test set by determining the mean square deviation of the predicted committor values from the true ones. Since for a large number of candidate variables an exhaustive test of all combinations of even only three or four of them using this procedure is impossible, a genetic algorithm is employed to search for the combination of variables that best predicts the committor, i.e., that yields the smallest mean square deviation.

The GNN-method can be used to search efficiently through very large pools of possible reaction coordinates. Computationally, the most expensive part of the procedure is the calculation of the committor values for a sufficiently large database; the cost for the training of the neural networks and the genetic optimization are relatively low. Ma and Dinner have applied the method to investigate the nature of the reaction coordinate for the isomerization of alanine dipeptide in vacuum and explicit solvent [28]. Their analysis comprised thousands of collective variables determined for each of the more than one thousand entries of the database. The collective variables included internal degrees of freedom of the dialanine molecule as well as solvent degrees of freedom. The genetic algorithm found that a combination of three variables, a solute dihedral angle, a solute–solute distance and a solvent generated electrostatic torque, is sufficient to parametrize the committor and, hence, specify transition states. The previously unknown role of long-ranged electrostatic interactions in this particular isomerization reaction was confirmed by the calculation of appropriate committor distributions.

8.5 Likelihood Maximization

As in the GNN-approach discussed in the previous section, the likelihood maximization method of Peters and Trout [29] determines the optimum reaction coordinate by screening a possibly large set of collective variables and finding the combination of collective variables that best fits the observed data. But in contrast to the GNN-method, the maximum likelihood approach does not require calculation of commitment probabilities. Rather, it builds on information about accepted and rejected shooting moves accumulated in a TPS simulation. This information is then analyzed using maximum likelihood estimation (MLE), a method of statistical analysis to determine the parameters of a postulated underlying model from a given finite set of data [276]. The central principle of this method is to find those model parameters that maximize the likelihood to observe the particular data set. In this sense, MLE looks for the most plausible explanation of the observations.

In order to apply this type of analysis to pathways generated in a TPS simulation, the shooting algorithm described in Sect. 5.2 has to be slightly modified. Since in this method one would like the acceptance probability of a shooting move attempted from r to be governed by $P(\mathrm{TP}|r)$, the new momenta at r need to be drawn from the Maxwell–Boltzmann distribution rather than obtained by small perturbation of the old momenta. This so-called aimless shooting algorithm differs from standard shooting in another important point: the shooting points are selected from a small region around the previous shooting point rather than from the whole path. Since shooting points near transition states have a higher probability to lead to reactive trajectories, this procedure leads to a population of shooting points that is densest near transition states [29]. As the TPS simulation carried out in this way proceeds, shooting points are stored together with the information on whether the trajectories started from them where accepted or rejected (i.e., whether they were reactive or not). This typically very large database is then subjected to a maximum likelihood analysis.

Application of MLE requires the specification of the underlying model in the form of parameter-dependent probability distributions for the data. In the approach of Peters and Trout, the data consist of the observed acceptances and rejections for a large number of shooting points together with the corresponding values of M collective variables $q_1, \ldots, q_M$ calculated for the collected shooting points. Each attempted shooting move, whether it is accepted or not, is viewed as a particular realization of the process whose statistics is described by $P(\text{TP}|q)$, the probability to be on a transition path given a particular value q of the (at this point unknown) reaction coordinate. This function $P(\text{TP}|q)$ is the model, which depends on various parameters. To specify the model and the parameters in detail, one first needs to postulate a specific functional dependence of $P(\text{TP}|q)$ on the reaction coordinate q (while the q_i are the collective variables, here the symbol q without subscript denotes the reaction coordinate). For a good reaction coordinate, $P(\text{TP}|q)$ is a function peaked at the value of q corresponding to the transition state and it decays to zero away from the peak [27]. The functional dependence chosen by Peters and Trout [29],

$$P(\text{TP}|q) = p_0 \left[1 - \tanh^2(q)\right], \tag{97}$$

is of this general form. Definition of the model is completed by stipulating how the reaction coordinate q depends on the M collective variables $q_1, \ldots, q_M$. A possibility to do that is

$$q = \alpha_0 + \sum_{k=1}^{M} \alpha_k q_k + \sum_{k,l=1}^{M} A_{kl} q_k q_l, \tag{98}$$

but depending on the particular situation other definitions might be more appropriate. On the basis of these definitions one can now construct a likelihood function, which quantifies the probability of the observed data as a function of the model parameters:

$$L(\alpha) = \prod_{r \in \text{acc}} P(\text{TP}|q(r)) \prod_{r \in \text{rej}} \left[1 - P(\text{TP}|q(r))\right]. \tag{99}$$

Here, α denotes all model parameters including p_0, the coefficients α_i and the matrix elements A_{ik}. The products extend over all accepted and rejected shooting points, respectively, and the dependence on the collective variables and the parameters α has been dropped for clarity on the right-hand side. Maximizing the likelihood function $L(\alpha)$ (or its logarithm) with respect to the parameters α yields the best reaction coordinate within the class of reaction coordinates permitted by the model.

Since it does not require the calculation of expensive committor histograms, likelihood maximization is a flexible and computationally very efficient method of finding reaction coordinates. Once a data set of shooting points with acceptance or rejection has been compiled, extension of the set of collective variables or the underlying model does not require any substantial additional computational effort. Algorithms are also available for screening large sets of collective variables as possible contributors to the reaction coordinate [29]. Some improvements of the maximum likelihood methods as well as a comparison with the GNN method of Ma and Dinner [28] are provided in [277]. MLE has been used to identify the mechanistic details of nucleation in the Ising model [29] and of structural solid–solid transitions of terephthalic acid [202].

9 Transition Path Theory and the String Method

As discussed in previous sections, the committor is the ideal reaction coordinate in the sense that it exactly quantifies how far a reaction has proceeded. This concept also provides the basis for transition path theory (TPT) [33, 278], a probabilistic framework developed by Vanden-Eijnden and collaborators to study the statistical properties of rare event trajectories. In TPT, isocommittor surfaces, i.e., surfaces on which all points have the same committor value, play a prominent role. Trajectories initiated from any point of an isocommittor surface have the same probability to reach the final rather than the initial state first. It can be shown [31] that the distribution of points where reactive trajectories pierce a given isocommittor surface is identical to the equilibrium distribution confined to that surface. From the committor and the equilibrium distribution one can determine the distribution of reactive trajectories, so-called reaction tubes, which contain entire reaction pathways with high probability, as well as the reaction rates, providing useful statistical information about the reaction mechanism.

One particular strength of TPT is that it provides a way to identify isocommittor surfaces directly without the need to generate dynamical trajectories by integration of the equations of motion, as is for instance done in several approaches described in the previous sections of this article. As discussed in [31, 33, 278], the committor function can be determined in principle by solving the backward Kolmogorov equation [279]. While this partial differential equation cannot be solved numerically except for simple, low-dimensional systems, it provides a starting point for the derivation of approximate algorithms that can be implemented on a computer. In the following, we will briefly outline the zero and finite temperature string methods, practical numerical approaches which follow from this perspective. We refer the reader to [31, 33, 278] for a detailed exposition of TPT.

At low temperatures, the reaction tube will be very narrow and concentrated mainly around the points of highest population on the isocommittor surfaces. In the framework of TPT, this observation leads to the zero-temperature string method

[30, 280], a numerical technique designed to find minimum energy pathways. In the zero-temperature string method, which resembles the NEB method described in Sect. 2.2, a path connecting A and B is represented by a smooth curve φ, the string. The minimum energy pathway satisfies the condition

$$(\nabla V)_{\perp}(\varphi) = 0, \tag{100}$$

where $(\nabla V)_{\perp}$ is the component of the gradient ∇V of the potential energy normal to the string. The minimum energy pathway is the pathway one obtains by walking downhill from a saddle point in the steepest descent direction. For overdamped dynamics at low temperatures, minimum energy pathways are the most likely transition routes. Once the minimum energy pathway is known, reaction rate constants can be calculated via TST. Starting from an arbitrary string that connects A and B, this minimum energy pathway can be found by evolving the string dynamically in a steepest descent way using the forces $f_{\perp} = -(\nabla V)_{\perp}$ until the string has converged. In practice, this is done by discretizing the string and carrying out the steepest descent dynamics on the images of the string. To enforce a particular parametrization of the string, for instance one defined through the normalized arc length, an appropriate constraint must be added to the evolution equation. Periodically, the images are redistributed on the string to exactly impose the parametrization that is only approximately maintained by the constraint. The normal forces used in the string method to drive the string from its initial form towards the minimum energy pathway are identical to the normal forces applied in the NEB method [62]. The two methods, however, differ in how the discrete images along the string are prevented from sliding towards the stable states. While in the NEB-method this is done by introducing tangential spring forces that act in a way to maintain approximately equal spacing between the images, in the string method this is done by imposing a certain parametrization as described above. Recently, an improved version of the string method has been developed which does not evolve the string using the force projection normal to the string [281]. Rather, the full force is used and then the string is reparametrized (the same idea can also be applied to the NEB method). In addition to being simple, this algorithm is also more accurate and stable. The zero-temperature string method, which has been validated using the rearrangement of a small Lennard-Jones cluster, has been used to study the pathways for thermally induced switching of magnetic films [30].

At finite temperatures, the minimum energy pathway is, in general, not representative for the ensemble of possibly very different transition pathways. Nevertheless, transition pathways often remain concentrated in one or a few transition tubes. If transition pathways are localized in this way, the finite temperature string method is applicable [282, 283]. Although the finite temperature string method has been developed in a probabilistic setting for systems with Markovian stochastic dynamics, it may be applied also to deterministic dynamics such as the Newtonian time evolution mainly used in MD simulations. In this case, however, no rigorous justification of the method is available. The central assumption of this method is that the isocommittor surfaces can be approximated by hyperplanes at least locally in the transition tube. The finite temperature string method then exploits the fact that on the isocommittor surfaces the distribution of points lying on reactive trajectories is identical to the equilibrium distribution restricted on that surface. It then follows from a variational principle for the solutions of the backward Kolmogorov equation that the isocommittor hyperplanes are normal to a string $\varphi(\alpha)$, defined as the average position of the equilibrium distribution on the hyperplane. Numerically the string $\varphi(\alpha)$, which lies at the center of the transition tube, and the isocommittor surfaces can be found by a procedure which is iterated until self-consistency is reached, i.e., until the hyperplanes are normal to the string $\varphi(\alpha)$. To do that in practice, the string is discretized and a hyperplane is attached to each image of the string. Then, the average position on each hyperplane is determined by carrying out independent constrained simulations to sample the equilibrium distribution on the hyperplanes, for instance with the blue moon sampling method [44]. This calculation yields a new string and a new set of perpendicular hyperplanes. As in the zero temperature string method, a particular parametrization is enforced by reparametrizing the string appropriately. The procedure is then iterated until the string has converged. As a result of a finite temperature string calculation one obtains the string at the center of the reactive tube, which can be viewed as a smooth representative of high likelihood transition pathways and which contains important information on the mechanism, as well as the isocommittor surfaces, from which transition states and transition rate constants follow. The finite temperature string method has been demonstrated to be an effective method also for high-dimensional systems with complex potential energy landscapes [282, 283] and has been employed to study the isomerization of alanine dipeptide in implicit and explicit solvent [283].

To study rare transitions in very large systems, Vanden-Eijnden and coworkers have recently developed a version of the string method in collective variables [32]. Provided that this set of collective variables $q_1(r), q_2(r), \ldots, q_M(r)$ is sufficient to capture the essence of the transition mechanism in the sense that the committor can be expressed as a function of these variables only, $p_B(r) = p_B(q_1(r), q_2(r), \ldots, q_M(r))$, this method yields the minimum free energy pathway (MFEP) of the reaction and the isocommittor surfaces. The MFEP is the most likely transition path in the space of the collective variables. Since in this method one is interested only in the MFEP rather than in mapping out the whole free energy landscape, the number of collective variables can be very large without adversely affecting the efficiency of the method. In [284] the authors have used the string method to study the hydrophobic collapse of a hydrophobic chain in more than 100,000 collective variables representing the water density. Analysis of the MFEP indicates that the collapse occurs by hydrophobic dewetting as proposed earlier by ten Wolde and Chandler [285] and that the system is driven over the barrier by a collective solvent motion which does not involve the chain degrees of freedom.

The string method is particularly suitable for studying systems evolving stochastically in the overdamped limit. The reason is that in the finite temperature string method one makes the assumption that once the system has entered B coming from A, any transition from B back to A is statistically uncorrelated. To satisfy this requirement the stable states A and B need to be appropriately defined, possibly in phase space rather than only in configuration space as can be done for overdamped dynamics. Often, however, it is already difficult to describe the stable states using only configurational properties and a full phase space definition of the stable states is even more challenging. An example where this issue creates problems are chemical reactions with weak coupling to the solvent (i.e., the energy diffusion regime) [286]. In this case, energy dissipation after activation is very slow such that the molecule may oscillate back and forth between the stable geometries several times before it finally settles in one of the stable states. In the reactive flux method such correlated recrossings are accounted for in the transmission coefficient and also time correlation functions determined with TPS properly describe the energy diffusion regime. In the string method, however, recrossings have to be eliminated by a suitable definition of the stable states in phase space, often a very difficult task for systems in which inertial effects are important.

The perspective adopted in the string method differs from that of other methods such as TPS. In the latter, one considers dynamical trajectories parametrized by physical time, while in the string method pathways are parametrized in a way that is numerically advantageous. This change in perspective provides the basis for the development of computational methods to determine the statistical properties of the reaction process in terms of the committor. These methods, however, require some approximations such as the assumption of hyperplanar isocommittor surfaces. Also, the dynamical details of transition trajectories are lost in the statistical description of the transition process. How to combine the statistical approach embodied in the TPT framework with methods such as TPS that do not suffer from the above limitations is an interesting but challenging open problem.

10 Conclusion

In this article we have reviewed several methods that allow the computational study of processes in which rare events play an important role. Quite a few robust and efficient new methods, including metadynamics, the finite temperature string method and TPS, have only become available during the last couple of years. These techniques have shown great potential for allowing the numerical study of processes that were hitherto not feasible. Nevertheless, all of these algorithms are somehow limited in their range of applicability and in the type of information they yield. An important challenge for the future will be to develop new techniques that combine the strengths of the various methods in a complementary way. Only in this way molecular simulation will truly be able to bridge the enormous time-scale gap that lies between the microscopic and the macroscopic world.

Acknowledgments C.D. acknowledges useful discussions with Phillip Geissler, Harald Oberhofer, Michael Grünwald, and Eric Vanden-Eijnden, and support from the Austrian Science Fund (FWF) under Grant No. P17178-N02. P.G.B. acknowledges the "Stichting voor Fundamenteel

Onderzoek der Materie (FOM)," which is financially supported by the "Nederlandse Organisatie voor Wetenschappelijk Onderzoek (NWO)." The authors are thankful to Michael Grünwald and Bianca Mladek for a critical reading of the manuscript.

References

1. Hansen J-P, Löwen H (2002) In: Nielaba P, Mareschal M, Ciccotti G (eds) Bridging time scales: molecular simulations for the next decade. Springer, Berlin
2. Abrams C, Delle Site L, Kremer K (2002) In: Nielaba P, Mareschal M, Ciccotti G (eds) Bridging time scales: molecular simulations for the next decade. Springer, Berlin
3. Likos CN (2001) Phys Rep 348:267
4. Louis AA, Bolhuis PG, Hansen J-P, Meijer EJ (2000) Phys Rev Lett 85:2522
5. Jaffe C, Ross SD, Lo MW, Marsden J, Farrelly D, Uzer T (2002) Phys Rev Lett 89:011101
6. Eigen M, De Maeyer L (1959) Z Elektrochem 59:986
7. Geissler PL, Dellago C, Chandler D, Hutter J, Parrinello M (2001) Science 291:2121
8. Wilding N, Landau DP (2002) In: Nielaba P, Mareschal M, Ciccotti G (eds) Bridging time scales: molecular simulations for the next decade. Springer, Berlin
9. Frenkel D, Smit B (2002) Understanding molecular simulation, 2nd edn. Academic, San Diego
10. Kirkpatrick S, Gelatt CD, Vecchi MP (1983) Science 220:671
11. Gottwald D, Kahl G, Likos CN (2005) J Chem Phys 122:204503
12. Oganov AR, Glass CW (2006) J Chem Phys 124:244704
13. Meerbach E, Dittmer E, Horenko I, Schütte C (2006) In: Ferrario M, Ciccotti G, Binder K (eds) Computer simulations in condensed matter: from materials to chemical biology. Springer, Berlin
14. Deuflhard P, Huisinga W, Fischer A, Schütte C (2000) Lin Alg Appl 315:39
15. Wales DJ, Doye JPK (1997) J Chem Phys A 101:5111
16. Henkelman G, Johannesson G, Jónsson H (2000) In: Schwartz SD (ed) Progress on theoretical chemistry and physics. Kluwer, Dordrecht
17. Elber R, Ghosh A, Cárdenas A, Stern H (2004) Adv Chem Phys 126:123
18. Hummer G, Kevrikidis IG (2003) J Chem Phys 118:10762
19. Laio A, Parrinello M (2002) Proc Natl Acad Sci USA 99:12562
20. Voter AF, Montalenti F, Germann TC (2002) Annu Rev Mater Res 32:321
21. Dellago C, Bolhuis PG, Csajka FS, Chandler D (1998) J Chem Phys 108:1964
22. Dellago C, Bolhuis PG, Chandler D (1998) J Chem Phys 108:9263
23. Bolhuis PG, Dellago C, Chandler D (1998) Faraday Discuss 110:421
24. Pratt LR (1986) J Chem Phys 85:5045
25. Geissler PL, Dellago C, Chandler D (1999) J Phys Chem B 103:3706
26. Bolhuis PG, Dellago C, Chandler D (2000) Proc Natl Acad Sci USA 97:5877
27. Hummer G (2004) J Chem Phys 120:516
28. Ma A, Dinner AR (2005) J Phys Chem B 109:6769
29. Peters B, Trout BL (2006) J Chem Phys 125:054108
30. Weinan E, Ren W, Vanden-Eijnden E (2002) Phys Rev B 66:052301
31. Weinan E, Ren W, Vanden-Eijnden E (2005) Chem Phys Lett 413:242
32. Maragliano L, Fischer A, Vanden-Eijnden E, Ciccotti G (2006) J Chem Phys 125:024106
33. Weinan E, Vanden-Eijnden E (2006) J Stat Phys 123:503
34. Stillinger FH (1976) J Chem Phys 65:3968
35. Louis AA, Bolhuis PG, Hansen J-P (2000) Phys Rev E 62:7961
36. Flory PJ (1953) Principles of polymer chemistry. Cornell University Press, Ithaca
37. Wales DJ (2003) Energy landscapes. Cambridge University Press, Cambridge
38. Stillinger FH, Weber TA (1983) Phys Rev A 28:2408

39. Swendsen RH, Wang J-S (1986) Phys Rev Lett 57:2607
40. Sugita Y, Okamoto Y (1999) Chem Phys Lett 314:141
41. Earl DJ, Deem MW (2005) Phys Chem Chem Phys 7:3910
42. Torrie GM, Valleau JP (1977) J Comput Phys 23:187
43. Kirkwood J (1935) J Chem Phys 3:300
44. Carter EA, Ciccotti G, Hynes JT, Kapral R (1989) Chem Phys Lett 156:472
45. Darve E, Pohorille AJ (2001) J Chem Phys 115:9169
46. Press WH, Teukolsky SA, Vetterling WT, Flannery BP (1992) Numerical recipes in C. Cambridge University Press, Cambridge
47. Cerjan CJ, Miller WH (1981) J Chem Phys 75:2800
48. Doye JPK, Wales DJ (1997) Z Phys D 40:194
49. Munro LJ, Wales DJ (1999) Phys Rev B 59:3969
50. Mousseau N, Barkema G (1996) Phys Rev Lett 77:4358
51. Mousseau N, Barkema G (1998) Phys Rev E 57:2419
52. Vocks H, Chubynsky MV, Barkema GT, Mousseau N (2005) J Chem Phys 123:244707
53. Henkelman G, Jónsson H (1999) J Chem Phys 111:7010
54. Tanase-Nicola S, Kurchan J (2003) Phys Rev Lett 91:188302
55. Tailleur J, Kurchan J (2007) Nat Phys 3:203
56. Elber R, Karplus M (1987) Chem Phys Lett 139:375
57. Czerminski R, Elber R (1990) Int J Quantum Chem 24:167
58. Ulitzky A, Elber R (1990) J Chem Phys 92:1519
59. Fischer S, Karplus M (1992) Chem Phys Lett 194:252
60. Sevick EM, Bell AT, Theodorou DN (1993) J Chem Phys 98:3196
61. Gillilan RE, Wilson KR (1992) J Chem Phys 97:1757
62. Jonssón H, Mills G, Jacobsen KW (1998) In: Berne BJ, Ciccotti G, Coker DF (eds) Classical and quantum dynamics in condensed phase simulations. World Scientific, Singapore, p. 385
63. Ionova IV, Carter EA (1993) J Chem Phys 98:6377
64. Dewar MJS, Healy EF, Stewart JJP (1984) J Chem Soc, Farad Trans II 80:227
65. Henkelman G, Jónsson H (2000) J Chem Phys 113:9978
66. Chu J-W, Trout BL, Brooks BR (2003) J Chem Phys 119:12708
67. Henkelman G, Uberaga BP, Jónsson H (2000) J Chem Phys 113:9901
68. Mills G, Jónsson H (1994) Phys Rev Lett 72:1124
69. Villarba M, Jónsson H (1995) Surf Sci 324:35
70. Eichler A, Hafner J (1999) Surf Sci 435:58
71. Ciobica IM, van Santen RA (2002) J Phys Chem B 106:6200
72. Sorensen MR, Mishin Y, Voter AF (2000) Phys Rev B 62:3658
73. Mathews DH, Case DA (2006) J Mol Biol 357:1683
74. Goldstein H (1980) Classical mechanics. Addison-Wesley, Reading
75. Landau LD, Lifshitz EM (1984) Mechanics. Pergamon, Oxford
76. Elber R (2006) In: Ferrario M, Ciccotti G, K. Binder (eds) Computer simulations in condensed matter: from materials to chemical biology. Springer, Berlin
77. Whittaker ET (1964) Analytical dynamics, 4th edn. Cambridge University Press, Cambridge
78. Passerone D, Parrinello M (2001) Phys Rev Lett 87:108302
79. Passerone D, Ceccarelli M, Parrinello M (2003) J Chem Phys 118:2025
80. Cho AE, Doll JD, Freeman DL (1994) Chem Phys Lett 229:218
81. Olender R, Elber R (1996) J Chem Phys 105:9299
82. Elber R, Ghosh A, Cárdenas A (2002) In: Nielaba P, Mareschal M, Ciccotti G (eds) Bridging time scales: molecular simulations for the next decade. Springer, Berlin
83. Elber R, Ghosh A, Cárdenas A (2002) Acc Chem Res 35:396
84. Onsager L, Machlup S (1953) Phys Rev 91:1505
85. Eastman P, Gronbech-Jensen N, Doniach S (2001) J Chem Phys 114:3823
86. Zaloj V, Elber R (2000) Comput Phys Commun 128:118
87. Uitdehaag JCM, van der Veen BA, Dijkhuizen L, Elber R, Dijkstra BW (2001) Prot Struct Funct Gen 43:327
88. Siva K, Elber R (2003) Prot Struct Funct Gen 50:63

89. Ghosh A, Elber R, Scheraga H (2002) Proc Natl Acad Sci USA 99:10394
90. Cárdenas A, Elber R (2003) Proteins 51:245
91. Cárdenas A, Elber R (2003) Biophys J 85:2919
92. Kevrikidis IG, Gear CW, Hyman JM, Kevrekidis PG, Runborg O, Theodoropoulos K (2003) Comm Math Sci 14:715
93. Amait MA, Kevrekidis IG, Maroudas D (2007) Appl Phys Lett 90:171910
94. Sriraman S, Kevrekidis IG, Hummer G (2005) Phys Rev Lett 95:130603
95. Iannuzzi M, Laio A, Parrinello M (2003) Phys Rev Lett 90:238302
96. Laio A, Parrinello M (2006) In: Ferrario M, Ciccotti G, K. Binder (eds) Computer simulations in condensed matter: from materials to chemical biology. Springer, Berlin
97. Grubmüller H (1995) Phys Rev E 52:2893
98. Schulze BG, Grubmüller H, Evanseck JD (2000) J Am Chem Soc 122:8700
99. Müller EM, de Meijere A, Grubmüller H (2002) J Chem Phys 116:897
100. Huber T, Torda A, van Gunsteren W (1994) J Comput Aided Mol Des 8:695
101. Wang F, Landau D (2001) Phys Rev Lett 86:2050
102. Bussi G, Laio A, Parrinello M (2006) Phys Rev Lett 96:090601
103. Micheletti C, Laio A, Parrinello M (2004) Phys Rev Lett 92:170601
104. Car R, Parrinello M (1985) Phys Rev Lett 55:2471
105. Laio A, Rodriguez-Fortea A, Gervasio FL, Ceccarelli M, Parrinello M (2005) J Phys Chem B 109:6714
106. Piana S, Laio A (2007) J Phys Chem B 111:4553
107. Bussi G, Gervasio FL, Laio A, Parrinello M (2006) J Am Chem Soc 128:13435
108. Martoňák R, Laio A, Parrinello M (2003) Phys Rev Lett 90:75503
109. Raitieri P, Martoňák R, Parrinello M (2005) Angew Chem Int Ed 44:3769
110. Ceccarelli M, Mercuri F, Passerone D, Parrinello M (2005) J Phys Chem B 109:17094
111. Stirling A, Iannuzzi M, Laio A, Parrinello M (2004) Chem Phys Chem 5:1558
112. Ensing B, de Vivo M, Liu Z, Moore P, Klein ML (2006) Acc Chem Res 39:73
113. Ceccarelli M, Danelon C, Laio A, Parrinello M (2004) Biophys J 87:58
114. Gervasio FL, Laio A, Parrinello M (2005) J Am Chem Soc 127:2600
115. Hanggi P, Talkner P, Borkovec M (1990) Rev Mod Phys 62:2551
116. Pechukas P (1981) Ann Rev Phys Chem 32:15
117. Anderson JB (1995) Adv Chem Phys 91:381
118. Vanden-Eijnden E, Tal FA (2005) J Chem Phys 123:184103
119. Marcelin A (1995) Ann Chim Phys 3:158
120. Eyring H (1935) J Chem Phys 3:107
121. Horiuti J (1938) Bull Chem Soc Jpn 13:210
122. Eyring H, Polanyi M (1931) Z Phys Chem B 12:279
123. Wigner E (1938) Trans Faraday Soc 34:29
124. Dellago C, Bolhuis PG, Geissler PL (2006) In: Ferrario M, Ciccotti G, Binder K (eds) Computer simulations in condensed matter: from materials to chemical biology. Springer Lecture Notes in Physics. Springer, Berlin, p. 349
125. Chandler D (1978) J Chem Phys 68:2959
126. Anderson JB (1973) J Chem Phys 58:4684
127. Kassel LS (1928) J Phys Chem 32:225; Rice OK, Ramsperger HC (1927) J Am Chem Soc 49:1617; Rice OK, Ramsperger HC (1927) J Am Chem Soc 50:617; Marcus R, Rice OK (1951) J Phys Colloid Chem 55:894
128. Geissler PL, Dellago C, Chandler D (1999) Phys Chem Chem Phys 1:1317
129. Truhlar DG, Garrett B (1984) Ann Rev Phys Chem 35:159
130. Truhlar DG, Garrett BC, Klippenstein SJ (1996) J Phys Chem 100:31
131. Tucker SC (1995) In: Talkner P, Hnggi P (eds) New trends in Kramers' reaction rate theory. Kluwer, Dordrecht
132. Johannesson G, Jónsson H (2001) J Chem Phys 115:9644
133. Fernández-Ramos A, Miller JA, Klippenstein SJ, Truhlar DG (2006) Chem Rev 106:4518
134. Yamamoto T (1960) J Chem Phys 33:281
135. Keck JC (1962) Discuss Faraday Soc 33:173

136. Bennett CH (1977) In: Christofferson R (ed) Algorithms for Chemical Computations, ACS Symposium Series No. 46. American Chemical Society, Washington DC
137. Ruiz-Montero MJ, Frenkel D, Brey JJ (1997) Mol Phys 90:925
138. van Erp TS, Bolhuis PG (2005) J Comput Phys 205:157
139. Waalkens H, Burbanks A, Wiggins S (2005) J Phys A Math Gen 38:L759
140. Waalkens H, Burbanks A, Wiggins S (2004) J Phys A 37:L257
141. Waalkens H, Burbanks A, Wiggins S (2004) J Chem Phys 121:6207
142. Tal FA, Vanden-Eijnden E (2006) Nonlinearity 19:501
143. Bortz AB, Kalos MH, Lebowitz JL (1975) J Comput Phys 17:10
144. Gillespie DT (1977) J Comput Phys 22:2340
145. Henkelman G, Jónsson H (2001) J Chem Phys 115:9657
146. Shirts MR, Pande VS (2001) Phys Rev Lett 86:4983
147. Voter AF (1998) Phys Rev B 57:13985
148. Pande VS, Baker I, Chapman J, Elmer SP, Khaliq S, Larson SM, Rhee YM, Shirts MR, Snow CD, Sorin EJ, Zagrovic B (2003) Biopolymers 68:91
149. Kum O, Dickson BM, Stuart SJ, Uberuaga BP, Voter AF (2004) J Chem Phys 121:9808
150. Birner S, Kim J, Richie DA, Wilkins JW, Voter AF, Lenosky T (2001) Solid State Commun 120:279
151. Uberuaga BP, Voter AF, Sieber KK, Sholl DS (2003) Phys Rev Lett 91:105901
152. Uberuaga BP, Stuart SJ, Voter AF (2007) Phys Rev B 75:014301
153. Voter AF (1997) J Chem Phys 106:4665
154. Voter AF (1997) Phys Rev Lett 78:3908
155. Duan XM, Sun DY, Gong XG (2001) Comput Mater Sci 20:151
156. Hamelberg D, McCammon JA (2005) J Am Chem Soc 127:13778
157. Becker KE, Fichthorn KA (2006) J Chem Phys 125:184706
158. Voter AF, Chen SP (1987) Mater Res Soc Symp Proc 82:175
159. Sørensen MR, Voter AF (2000) J Chem Phys 112:9599
160. Montalenti F, Voter AF, Ferrando R (2002) Phys Rev B 66:205404
161. Montalenti F, Sørensen MR, Voter AF (2001) Phys Rev Lett 87:126101
162. Cogoni M, Mattoni A, Uberuaga BP, Voter AF, Colombo L (2005) Appl Phys Lett 87:191912
163. Sprague JA, Montalenti F, Uberuaga BP, Kress JD, Voter AF (2002) Phys Rev B 66:205415
164. Bolhuis PG, Chandler D, Dellago C, Geissler PL (2002) Ann Rev Phys Chem 53:291
165. Dellago C, Bolhuis PG, Geissler PL (2002) Adv Chem Phys 123:1
166. Dellago C, Chandler D (2002) In: Nielaba P, Mareschal M, Ciccotti G (eds) Molecular simulation for the next decade. Springer, Berlin, p. 321
167. Dellago C (2005) In: Yip S (ed) Handbook of materials modeling. Springer, Berlin, p. 1585
168. Dellago C (2007) In: Pohorille A, Chipot C (eds) Free energy calculations: theory and applications in chemistry and biology. Springer, Berlin
169. Dellago C, Bolhuis PG (2007) In: Reiher M (ed) Topics in Current Chemistry, vol. 268. Springer, Berlin, p. 291
170. Dellago C, Geissler PL (2003) In: Proceedings of The Monte Carlo Method in the Physical Sciences: Celebrating the 50th anniversary of the Metropolis algorithm. AIP Conference Proceedings, vol. 690
171. Geissler PL, Dellago C (2004) J Phys Chem B 108:6667
172. Chandrasekhar S (1943) Rev Mod Phys 15:1
173. Zwanzig R (2001) Nonequilibrium statistical mechanics. Oxford University Press, Oxford
174. Metropolis N, Rosenbluth AW, Rosenbluth MN, Teller AH, Teller E (1953) J Chem Phys 21:1087
175. Dellago C, Bolhuis PG, Csajka FS, Chandler D (1999) J Chem Phys 110:6617
176. Geyer CJ, Thompson EA (1995) J Am Stat Soc 90:909
177. Vlugt TJH, Smit B (2001) Phys Chem Comm 2:1
178. Bolhuis PG (2003) J Phys Cond Matter 15:S113
179. Grünwald M, Geissler PL, Dellago C (2008), J Chem Phys (in print)
180. Hu J, Ma A, Dinner AR (2006) J Chem Phys 125:114101
181. Grünwald M, Rabani E, Dellago C (2006) Phys Rev Lett 96:255701

182. Grünwald M, Geissler PL, Dellago C (2007) J Chem Phys 127:154718
183. Dellago C, Bolhuis PG (2004) Mol Sim 30:795
184. van Erp TS, Moroni D, Bolhuis PG (2003) J Chem Phys 118:7762
185. Juraszek J, Bolhuis PG (2006) Proc Natl Acad Sci USA 103:15859
186. Moroni D, Bolhuis PG, van Erp TS (2004) J Chem Phys 120:4055
187. van Erp TS (2007) Phys Rev Lett 98:268301
188. Bolhuis PG (2008) J. Chem. 129, 114108
189. Moroni D, van Erp TS, Bolhuis PG (2005) Phys Rev E 71:056709
190. Allen RJ, Warren PB, ten Wolde PR (2005) Phys Rev Lett 94:018104
191. Allen RJ, Frenkel D, ten Wolde PR (2006) J Chem Phys 124:024102
192. Allen RJ, Frenkel D, ten Wolde PR (2006) J Chem Phys 124:194111
193. Valeriani C, Allen RJ, Morelli MJ, Frenkel D, ten Wolde PR (2007) J Chem Phys 127:114109
194. Borrero EE, Escobedo FA (2006) J Chem Phys 125:164904
195. Faradjian T, Elber R (2004) J Chem Phys 120:10882
196. West AMA, Elber R, Shalloway D (2007) J Chem Phys 126:145104
197. Elber R (2007) Biophys J 92:L85
198. Auer S, Frenkel D (2005) Adv Poly Sci 173:149
199. Pan AC, Chandler D (2004) J Phys Chem B 108:19681
200. Moroni D, ten Wolde PR, Bolhuis PG (2005) Phys Rev Lett 94:235703
201. Zahn D (2007) J Phys Chem B 111:5249
202. Beckham GT, Peters B, Starbuck C, Variankaval N, Trout BL (2007) J Am Chem Soc 129:4714
203. Zahn D (2006) J Phys Chem B 110:19601
204. Zahn D, Grin Y, Leoni S (2005) Phys Rev B 72:064110
205. Zahn D (2004) Phys Rev Lett 93:227801
206. Zahn D, Leoni S (2006) J Phys Chem B 110:10873
207. Zahn D, Hochrein O, Leoni S (2005) Phys Rev B 72(9):094106
208. Zahn D (2004) J Solid State Chem 177:3590
209. Leoni S, Zahn D (2004) Z Kristall 219:339
210. Zahn D, Leoni S (2004) Z Kristall 219:345
211. Zahn D, Leoni S (2004) Phys Rev Lett 92:250201
212. Zahn D (2004) Phys Rev Lett 92:040801
213. Day TJF, Schmitt UW, Voth GA (2000) J Am Chem Soc 122:12027
214. Park JM, Laio A, Iannuzzi M, Parrinello M (2006) J Am Chem Soc 128:11318
215. Lo CS, Radhakrishnan R, Trout BL (2005) Catalaysis Today 105:93
216. Snee PT, Shanoski J, Harris CB (2005) J Am Chem Soc 127:1286
217. Zahn D (2004) Chem Phys 300:79
218. Zahn D, Seifert G (2004) J Phys Chem B 108:16495
219. Ensing B, Baerends EJ (2002) J Phys Chem A 106:7902
220. Pagliai M, Raugei S, Cardini G, Schettino V (2002) J Chem Phys 117:2199
221. Marti J, Csajka FS, Chandler D (2000) Chem Phys Lett 328:169
222. Laria D, Rodriguez J, Dellago C, Chandler D (2001) J Phys Chem A 105:2646
223. Lee JY (2004) Chem Phys 299:123
224. Csajka FS, Chandler D (1998) J Chem Phys 109:1125
225. Marti J, Csajka FS (2000) J Chem Phys 113:1154
226. Marti J (2001) Mol Sim 27:169
227. Bolhuis PG, Chandler D (2000) J Chem Phys 113:8154
228. Vlugt TJH, Dellago C, Smit B (2000) J Chem Phys 113:8791
229. Pool R, Bolhuis PG (2007) J Chem Phys 126:244703
230. Ramirez J, Laso M (2001) J Chem Phys 115:7285
231. Radhakrishnan R, Schlick T (2004) Proc Natl Acad Sci USA 101:5970
232. Radhakrishnan R, Schlick T (2004) J Chem Phys 121:2436
233. McCormick TA, Chandler D (2003) J Phys Chem B 107:2796
234. Hagan MF, Dinner AR, Chandler D, Chakraborty AK (2003) Proc Natl Acad Sci USA 100:13922

235. Crehuet R, Field MJ (2007) J Phys Chem B 111:5708
236. ten Wolde PR, Chandler D (2002) Proc Natl Acad Sci USA 99:6539
237. Bolhuis PG (2003) Proc Natl Acad Sci USA 100:12129
238. Bolhuis PG (2005) Biophys J 88:50
239. Radhakrishnan R, Yang LJ, Arora K, Schlick T (2004) Biophys J 86:34A
240. Radhakrishnan R, Schlick T (2005) J Am Chem Soc 127:13245
241. Basner JE, Schwartz SD (2005) J Am Chem Soc 127:13822
242. Marti J, Csajka FS (2004) Phys Rev E 69:061918
243. Marti J (2004) J Phys Condens Matter 16:5669
244. Ma L, Cui Q (2007) J Am Chem Soc 129:10261
245. Wang YL, Schlick T (2007) BMC Struct Biol 7:7
246. Merolle M, Garrahan JP, Chandler D (2005) Proc Natl Acad Sci USA 102:10837
247. Crooks GE, Chandler D (2001) Phys Rev E 64:026109
248. Geissler PL, Chandler D (2000) J Chem Phys 113:9759
249. Sun SX (2003) J Chem Phys 118:5769
250. Ytreberg FM, Zuckerman DM (2004) J Chem Phys 120:10876
251. Oberhofer H, Dellago C, Geissler PL (2005) J Phys Chem B 69:6902–6915
252. Lechner W, Dellago C (2007) J Stat Mech Theor Exp P04001
253. Athènes M (2004) Eur Phys J B 38:651
254. Oberhofer H, Dellago C (2008) Comput Phys Commun 179:41
255. Imparato A, Peliti L (2007) J Stat Mech L02001
256. Imparato A, Peliti L (2007) Compt Rend Physique 8:556
257. Wales DJ (2002) Mol Phys 100:3285
258. Wales DJ (2005) Phys Biol 2:S86
259. Trygubenko SA, Wales DJ (2006) Mol Phys 104:1497
260. Evans DA, Wales DJ (2004) J Chem Phys 121:1080
261. Best RB, Hummer G (2005) Proc Natl Acad Sci USA 102:6732
262. Onsager L (1938) Phys Rev 54:554
263. Du R, Pande V, Grosberg AY, Tanaka T, Shakhnovich EI (1998) J Chem Phys 108:334
264. Gasser U, Weeks ER, Schofield A, Pusey PN, Weitz DA (2001) Science 292:258
265. Ryter D (1987) Physica A 142:103
266. Berezhovski A, Szabo A (2006) J Chem Phys 125:104902
267. Klosek MM, Matkowsky BJ, Schuss Z (1991) Ber Bunsenges Phys Chem 95:331
268. Pollak E, Berezhkovskii AM, Schuss Z (1994) J Chem Phys 100:334
269. Talkner P (1994) Chem Phys 180:199
270. Berezhovski A, Szabo A (2005) J Chem Phys 122:014503
271. Metzner P, Schütte C, Vanden-Eijnden E (2006) J Chem Phys 125:084110
272. Peters B (2006) J Chem Phys 125:241101
273. Quaytman SL, Schwartz SD (2007) Proc Natl Acad Sci USA 104:12253
274. So S-S, Karplus M (1996) J Med Chem 39:1521
275. Dinner A, So S-S, Karplus M (2002) Adv Chem Phys 120:1
276. Edwards AWF (1972) Likelihood. Cambridge University Press, Cambridge
277. Peters B, Beckham GT, Trout BL (2007) J Chem Phys 127:034109
278. Vanden-Eijnden E (2006) In: Ferrario M, Ciccotti G, Binder K (eds) Computer simulations in condensed matter: from materials to chemical biology, Springer Lecture Notes in Physics. Springer, Berlin, p. 453
279. Gardiner CW (1985) Handbook of stochastic methods. Springer, Berlin
280. Ren W (2003) Comm Math Sci 1:377
281. Weinan E, Ren W, Vanden-Eijnden E (2007) J Chem Phys 126:164103
282. Weinan E, Ren W, Vanden-Eijnden E (2005) J Phys Chem B 109:6688
283. Ren W, Vanden-Eijnden E, Maragakis P, Weinan E (2005) J Chem Phys 123 134109
284. Miller TF III, Vanden-Eijnden E, Chandler D (2007) Proc Natl Acad Sci USA 104:14559
285. ten Wolde P-R, Chandler D (2002) Proc Natl Acad Sci USA 99:6539
286. Wilson MA, Chandler D (1990) Chem Phys 149:11

Index